# MEDICINAL PLANTS OF BORNEO

## Natural Products Chemistry of Global Plants

Editor: Raymond Cooper

This unique book series focuses on the natural products chemistry of botanical medicines from different countries such as Sri Lanka, Cambodia, Brazil, China, Africa, Borneo, Thailand, and Silk Road Countries. These fascinating volumes are written by experts from their respective countries. The series will focus on the pharmacognosy, covering recognized areas rich in folklore as well as botanical medicinal uses as a platform to present the natural products and organic chemistry. Where possible, the authors will link these molecules to pharmacological modes of action. The series intends to trace a route through history from ancient civilizations to the modern day showing the importance to man of natural products in medicines, foods, and a variety of other ways.

Recent Titles in This Series

Medicinal Plants of Borneo
*Simon Gibbons and Stephen P. Teo*

Natural Products and Botanical
Medicines of Iran
*Reza Eddin Owfi*

Natural Products of Silk Road
Plants
*Raymond Cooper and Jeffrey
John Deakin*

Brazilian Medicinal Plants
*Luzia Modolo and Mary Ann Foglio*

Medicinal Plants of Bangladesh
and West Bengal: Botany, Natural
Products, and Ethnopharmacology
*Christophe Wiart*

Traditional Herbal Remedies of Sri Lanka
*Viduranga Y. Waisundara*

# MEDICINAL PLANTS OF BORNEO

Edited by

Simon Gibbons and Stephen P. Teo

CRC Press is an imprint of the
Taylor & Francis Group, an **informa** business

First edition published 2021
by CRC Press
6000 Broken Sound Parkway NW, Suite 300, Boca Raton, FL 33487-2742

and by CRC Press
2 Park Square, Milton Park, Abingdon, Oxon, OX14 4RN



*Library of Congress Cataloging-in-Publication Data*
Names: Gibbons, Simon (Professor of medicinal phytochemistry), editor. | Teo, Stephen P., editor.
Title: Medicinal plants of Borneo / edited by Simon Gibbons and Stephen P. Teo.
Description: 1st edition. | Boca Raton : CRC Press, 2021. | Series: Natural products chemistry of global plants | Includes bibliographical references and index.
Identifiers: LCCN 2020050287 (print) | LCCN 2020050288 (ebook) | ISBN 9781138601079 (paperback) | ISBN 9780367758752 (hardback) | ISBN 9780429470332 (ebook)
Subjects: LCSH: Materia medica, Vegetable--Borneo. | Medicinal plants--Borneo. | Pharmaceutical chemistry.
Classification: LCC RS180.B67 M43 2021 (print) | LCC RS180.B67 (ebook) | DDC 615.3/21095983--dc23
LC record available at https://lccn.loc.gov/2020050287
LC ebook record available at https://lccn.loc.gov/2020050288

ISBN: 978-0-367-75875-2 (hbk)
ISBN: 978-1-138-60107-9 (pbk)
ISBN: 978-0-429-47033-2 (ebk)

Typeset in Times New Roman
by MPS Limited, Dehradun

# Contents

# Series Preface

## NATURAL PRODUCTS CHEMISTRY OF GLOBAL PLANTS

CRC Press is publishing a new book series on the *Natural Products Chemistry of Global Plants*. This new series will focus on pharmacognosy covering uses in folklore, presenting natural products and, where possible, linking these to pharmacological modes of action. The book series on botanical medicines from different countries includes, but is not limited to, Bangladesh, Borneo, Brazil, Cambodia, Cameroon, Ecuador, Iran, Madagascar, S. Africa, Sri Lanka, The Silk Road, Thailand, Turkey, Uganda, Vietnam and Yunnan Province (China), and is being written by experts from each country. The intention is to provide a platform to bring forward information from under-represented regions.

Medicinal plants are an important part of human history, culture and tradition. Plants have been used for medicinal purposes for thousands of years. Anecdotal and traditional wisdom concerning the use of botanical compounds is documented in the rich histories of traditional medicines. Many medicinal plants, spices and perfumes changed the world through their impact on civilization, trade and conquest. Folk medicine is commonly characterized by the application of simple indigenous remedies. People who use traditional remedies may not understand in our terms the scientific rationale for why they work but know from personal experience that some plants can be highly effective.

This series provides rich sources of information from each region. An intention of the series of books is to trace a route through history from ancient civilizations to the modern day showing the important value to humankind of natural products in medicines, in foods and in many other ways. Many of the extracts are today associated with important drugs, nutrition products, beverages, perfumes, cosmetics and pigments, which will be highlighted.

The books will be written for both chemistry students who are at university level and for scholars wishing to broaden their knowledge in pharmacognosy. Through examples of chosen herbs and plants, the series will describe the key natural products and their extracts with emphasis upon sources, an appreciation of these complex molecules and applications in science.

In this series the chemistry and structure of many substances from each region will be presented and explored. Often books describing folklore medicine do not describe the rich chemistry or the complexity of the natural products and their respective biosynthetic building blocks. The story becomes more fascinating by drawing upon the chemistry of functional groups to show how they influence the chemical behavior of the building blocks which make up large and complex natural products. Where possible, it will be advantageous to describe the pharmacological nature of these natural products.

**R Cooper PhD, Editor-in-Chief**
*Department of Applied Biology and Chemical Technology*
*The Hong Kong Polytechnic University Hong Kong*

# Preface

It is a pleasure to write a preface for this exciting book on the pharmacognosy of Bornean plants. This is the first on the medicinal plants of the whole island of Borneo; whilst existing works are mainly piecemeal and focus on a particular ethnic group or region within the island.

The list of species furnished in this text is by no means exhaustive and the major and more commonly used plants are highlighted, although rare ones are also occasionally included. Many medicinal plants can of course have more than one biological activity. For example, *Tinospora crispa* shows activities against malaria and diabetes. It must be noted however, that the majority of plant-based medicines are little studied in terms of their chemistry and pharmacology, and in general, their toxicities are even less well understood. Therefore, a word of caution must be given to the reader – some of these species, despite having an extensive history of being able to treat illnesses, can be toxic and must only be used with extreme care. This book is not a substitute for any medical advice.

Borneo, located in Southeast Asia, is one of the global hotspots for biodiversity and is home to a highly diverse array of biota. Coupled with a diversity of ethnic groups, both indigenous and migrants, many species within this varied biota find their uses in Ethnobotany, especially as medicines. This book is a documentation of some of the wealth of this Pharmacognosy that is used in both folk medicine in the different regions of Borneo (Malaysia, Indonesia and Brunei), but also those with bioactive compounds against diseases, identified through modern techniques in the screening of plant extracts. Apart from these uses, this book also provides information such as their Phytochemistry, including chemical structures of the bioactive compounds as well as their respective pharmacological activities. The book is divided into various chapters with plants of similar usage grouped together under the same chapter.

Dr Stephen Teo and the other contributors are to be congratulated on providing a highly informative additional volume to our knowledge of global natural products and the medicinal plants of Borneo.

**Simon Gibbons FRSC FLS**

*Professor of Natural Product Chemistry*

*Head of School, School of Pharmacy, University of East Anglia*

# Acknowledgments

This book would not have materialized if it were not for the assistance and cooperation of many people. Grateful thanks are due to the many contributors and editors for putting together all the materials for the book. Different authors contributed photographs and are all hereby gratefully acknowledged. Last but far from least, our gratitude is to the team at CRC Press, including Ms. Hilary Lafoe and Ms. Jessica Poile and the editor-in-chief, Dr Ray Cooper, for initiating the publication of this timely and useful book.

# Editors

**Simon Gibbons** is currently Head of School, School of Pharmacy, University of East Anglia. Among his research interests are antibacterial plant natural products, natural product bacterial resistance modifying agents, the chemistry and pharmacology of novel psychoactive substances and drugs of abuse and phytochemistry of herbal drugs. He is founding Editor-in-Chief of the journal *Phytochemistry Letters*, and currently serves on the Editorial Advisory Board of the book series Progress in the Chemistry of Organic Natural Products ("Zechmeister"; Springer Verlag, Vienna) and is a member of the Editorial Boards of the journals *Natural Product Reports, Planta Medica, Phytochemistry Reviews, Phytochemical Analysis, Phytotherapy Research, Fitoterapia, Pharmaceutica Scientia* and *Chinese Journal of Natural Medicine.*

**Stephen P. Teo** is attached to the Forest Department Sarawak and has more than two decades of experience in the department. His interest is in the plant biodiversity of Borneo, and he has published papers and/or books on plant taxonomy, ecology and conservation, economic plants, pharmacognosy and phytochemistry as well as undertaking extensive field work throughout Borneo.

# Contributors

**Stephen P. Teo** is attached to the Forest Department Sarawak and has more than two decades of experience working in the department. His interest is in the plant biodiversity of Borneo and has published papers and or books on plant taxonomy, ecology and conservation, economic plants, pharmacognosy and phytochemistry as well as undertaking extensive field work throughout Borneo.

Affiliation: Forest Department Sarawak, Malaysia

**Farid Kuswantoro** is currently working as a research assistant at the Bali Botanic Garden, Indonesian Institute of Sciences (LIPI). He started working in the botanic garden in 2015 as a research candidate and attained his current position in December 2017. His first contact with Borneo and its plant species came in 2014 when the Bogor Botanic Garden assigned him to assist the development of the newly established Balikpapan Botanic Garden in East Kalimantan.

Affiliation: Eka Karya, Bali Botanic Garden, Indonesian Institute of Sciences (LIPI), Formerly with Balipapan Botanic Garden, East Kalimantan, Indonesia

**Irawan Wijaya Kusuma** is a Lecturer in the field of forest products chemistry at the Laboratory for Forest Products Chemistry and Renewable Energy, Faculty of Forestry, Mulawarman University, Indonesia. He graduated from Mulawarman University, Indonesia for his undergraduate and masters degrees and obtained his PhD from Ehime University, Japan. His works are related to the exploration of tropical plants having biological activities and the discovery of antimicrobial, antioxidant and enzyme inhibitory compounds from plant resources. He has published more than 70 articles in peer-reviewed publications, and written 3 books as well as been granted 4 patents.

Affiliation: Laboratory for Forest Products Chemistry and Renewable Energy, Faculty of Forestry, Universitas Mulawarman, East Kalimantan, Indonesia

**Tukirin Pratomidhardjo** was attached to the Indonesian Institute of Sciences (LIPI). Bogor, Indonesia and retired as a professor in 2017. Since 2014 and up to now, he leads the Indonesian Forum for Threatened Trees (FPLI). His interests are in botany and plant ecology, and he has conducted field and research work throughout Indonesia including in Borneo.

Affiliation: Formerly with the Indonesian Institute of Sciences (LIPI). Bogor, Indonesia

**Meekiong Kalu** is a botanist who lectures at the Universiti Malaysia Sarawak (UNIMAS), Kota Samarahan, Sarawak. He has special interests in the Order Zingiberales (particularly Musaceae and Zingiberaceae) as well as the Dipterocarpaceae family. He has worked on the medicinal plants of Sarawak since 2003, mainly on the documentation of traditional knowledge.

Affiliation: Faculty of Rescource Science, Universiti Malaysia Sarawak, Malaysia

**Mohd Razip Asaruddin** is an Associate Professor in chemistry and also the deputy dean of the Faculty of Resource Science and Technology (Industry and Community Engagement), Universiti Malaysia Sarawak, Kota Samarahan, Malaysia. He holds an MPharm. Sc. from Kyoto University and a doctorate in drug design and pharmaceutic technology from Universiti Sains Malaysia. His interests are in organic chemistry, natural products chemistry and drug design.

Affiliation: Faculty of Resource Science, Universiti Malaysia Sarawak, Malaysia

# Aims and Purposes

This book, *Medicinal Plants of Borneo,* is meant to acquaint the general public and lay people with this important resources of medicinal plants found in Borneo. Traditional ethnomedicine continues to be widely practiced in Borneo. Prohibitive costs and access to treatments, side effects of synthetic drugs and problems with resistance to existing anti-microbial drugs have resulted in a surge in the use of plant materials as a source of medicines for a great array of human illness.

Medicinal plants are still an indispensable reservoir of pharmaceutical products and novel natural products are still the best sources of novel bioactive compounds. It is estimated that only 15% of higher plants have been investigated for potentially useful biological activity. A quarter of present medicine comes from natural products while another quarter is derived from natural sources as lead compounds.

In the last two decades, there has been an increase in the use of herbal medicine but research data in this field remain meagre. To this end, it is the intention of this publication to highlight the existing useful information on selected medicinal plants associated with Borneo.

## A GUIDE TO THE BOOK

This book focuses on the scientific and ethnomedical aspects of the medicinal plants used by the different ethnic groups throughout Borneo. The tropical rainforests of Borneo are an important source of such medicinal plants but these are also supplemented by other plants that have been introduced either by migrants or grown as ornamental plants but have now been used as medicinal plants. This book aims to introduce some of the more familiar as well as some of the rarer examples of medicinal plants together with their chemistry and the existing gaps in their research work.

Whilst we know a lot about the background of many of these medicinal plants used in Borneo, their chemistry, pharmacology and in some cases clinical effectiveness are often not mentioned or discussed and hard to come by in available books. Along with their traditional uses, this guide presents up-to-date information on botanical, pharmacological and chemical properties of the plants as well as photographs of most of the plants mentioned in the book.

The book is divided into various chapters based on plants activities or constituents with activities on malaria, bacteria, fungi, wound healing and so forth. This book also presents the current scientific evidence of studies done both *in vivo* and *in vitro* for medicinal plants.

Many medicinal plants are often combined or mixed with various other medicinal plants or herbs but the focus here is on single medicinal plants. The medicinal plants selected encompass a wide range from ferns to flowering plants and include different plant parts such as leaves, fruits, stems and so forth.

Medicinal plants, especially those from the tropical rainforests, need to be more widely cultivated and their traditional uses examined in a scientific and rigorous manner in order that their biomolecular activities can be confirmed and further translated into clinical uses and benefits.

It is the aim of this book to not only contribute to the enhancement of basic knowledge used in traditional phytomedicine in Borneo but will also provide pointers to the gaps in research work.

# Introduction

*Stephen P. Teo*

## BACKGROUND

Borneo is the world's third-largest island and the second largest tropical island after New Guinea. It occupies a central position in Southeast Asia and the island lies between 07° 4′N and 04° 7′S and between 114° 50′E and 118° 56′E with the equator running approximately midway across the island. It has an area of approximately 743,330 km$^2$ (287,000 sq. miles) and is shared by 3 countries – Indonesia (provinces of East, West, North, South and Central Kalimantan), Malaysia (Sabah and Sarawak) as well as the Sultanate of Brunei (Figure I.1).

## PHYSICAL SETTING

Borneo is relatively flat with extensive peat swamps along the coastal areas but with a mountainous terrain in the central part of the island that serves as a political divide between Sarawak, Malaysia and Indonesian Borneo (Figure I.1). This extends into Sabah, Malaysia and reaches its peak on Mount Kinabalu at 4,095 m (13,435 ft), which is the highest mountain in Borneo and Southeast Asia. Borneo is also drained by many great rivers that start from the mountainous interior and flow toward the seas. Among the longest rivers are the Kapuas (West Kalimantan – 1143 km), Mahakam (East Kalimantan – 980 km), Barito (Central Kalimantan – 900 km), Rejang (Sarawak – 563 km) and Kinabatangan (Sabah – 560 km) rivers.

Borneo has an equatorial type of climate. Rainfall is significant throughout the year and varies from about 150 cm to over 400 cm with an average of 250 cm per year. Often, there are no signs of distinct wet and dry seasons. The difference in precipitation between the driest month and the wettest month is about 40 cm. The weather is humid with relative humidity ranges from 70% to 85%. Temperature varies between 20°C and 32°C along the coastal areas but is cooler at night in the interior and at higher elevation.

## ETHNIC DIVERSITY OF BORNEO

Borneo is sparsely populated and has about 21.3 million inhabitants (in 2014), a population density of 29 inhabitants per km$^2$ (75 inhabitants per sq. mile). Most of the population are found in coastal cities with the remainder scattered in small towns and villages along the rivers. The population consists mainly of the over 50 indigenous Dayak ethnic groups (Bidayuh, Embaloh, Iban, KadazanDusun, Kayan, Kenyah, Lun Bawang, Lun Dayeh, Melanau, Murut, Penan, Rungus and so forth), Malay (Bajau, Banjarese, Belait, Bruneian, and so forth), and Javanese and Chinese, among others.

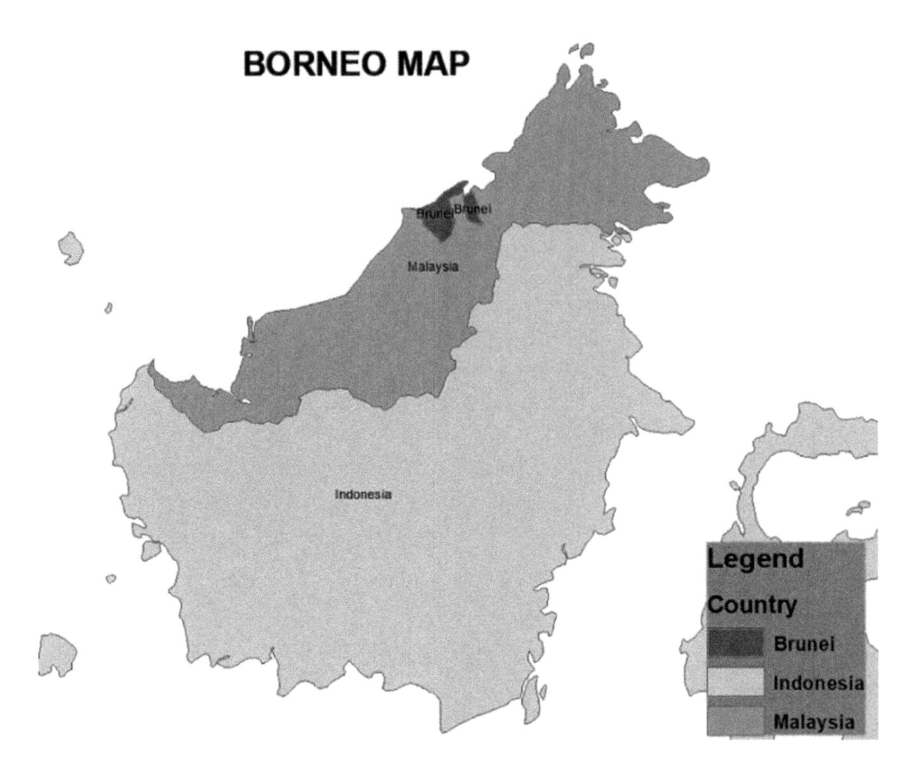


FIGURE I.1 Map of Borneo (Malaysia, Brunei and Indonesia).

## PLANT BIODIVERSITY OF BORNEO

Borneo is located in the ethnobotanical region of Malesia and occupies the eastern portion known collectively as Sundaland, which are made up of Sumatra, Peninsular Malaysia and Pahlawan Island in the Philippines apart from Borneo and are floristically similar as opposed to Sahul comprising the northern coast of Australia, New Guinea, Aru and neighbouring islands. The Sunda shelf and Sahul shelf are separated by a deep trench which is thought to prevent migration of plants and animals. The Borneo rainforests are some of the oldest in the world and also some of the most biodiverse due to past climatic and geological histories. Borneo's tropical rainforests and climate provide the ideal conditions for a wide variety of species to thrive. There are about 15,000 species of flowering plants with approximately 3,000 species of trees (267 species are dipterocarps). They are also home to thousands of non-flowering plants, lichens and fungi. It is also the centre of evolution and radiation of many endemic species of plants and animals.

There are seven distinct ecosystems in Borneo. The Borneo biodiverse lowland mixed dipterocarp rainforests cover most of the island. Other significant lowland ecosystems are peat swamp forests, heath or kerangas/kerapah forests, forest over limestone hills and ultramafic outcrops (particularly those in Sabah and South Kalimantan) and finally montane forests and mangrove forests The Borneo montane rainforests lie mainly in the central highlands of the island, above the 1,000 meters elevation. The highest elevations of Mount Kinabalu include mountain alpine meadow and shrubland, rich in endemic species.

## HISTORICAL AND TRADITIONAL USES OF MEDICINAL PLANTS OF BORNEO

In ancient times, Borneo was, for centuries, one of the destinations for traders from China, India, Persia, the Middle East and other Southeast Asian countries (Sellato, 2002). These barter traders were mostly attracted by the non-timber forest products such as edible bird's nest, ivory, rattan, incense wood (gaharu), spices, dye stuffs, fibre, tree barks, roots, oil (e.g. from Sindora and Dipterocarpus) and also medicinal plants such as dragon blood resins from *Daemonorops draco* and camphor from *Dryobalanops aromatica.* Durable materials, however, like resins such as Buseraceous resin, dammar, gutta-percha and dragon's blood have been found in ancient shipwrecks in the waters of Southeast Asia and off Borneo which dated back to as early as the 12th Century with origin in southern China Fujian Province (Pappas, 2018).

The various ethnic groups as well as the great plant diversity contribute greatly to the diversity in the usage of plants for medicinal purposes. This history has led to multiple uses of plants for various purposes including medicinal plants in folk medicine.

## CAMPHOR

In ancient times, camphor particularly attracted Arab traders to Borneo. Camphor is an aromatic, crystalline substance with a chemical structure that exists in 2 enantiomers, (*R*)- (**1**) and (*S*)-camphor (**2**), as shown in Figure I.2. The camphor isomer obtained from *Dryobalanops aromatica* is dextrorotatory; the levorotatory form is obtainable from other species of plants. The camphor can be collected from fissures or cavities in the trunks of the *Dryobalanops aromatica* tree found in Sumatra, Peninsular Malaysia and Borneo: while a clear yellow liquid known as 'oil of camphor' is obtained by tapping young trees or by distillation of its wood, where camphor sometimes crystalizes in older specimens. Both crystals and oil are used in capsules, infusions, lotions, pills, powders, and rubbing oils.

Camphor is used for its scent, and for manufacture of its esters, as an ingredient in cooking (mainly in India), as an embalming fluid, for medicinal purposes, and in religious ceremonies. For medicinal purposes, camphor is used internally for

**FIGURE I.2** The 2 enantiomers of camphor (*R*)- (**1**) and (*S*)-camphor (**2**).

resuscitation after fainting and used in convulsions associated with high fever, cholera, and pneumonia. It is applied externally for ringworm, rheumatism, abscesses, boils, cold sores, mouth ulcers, sore throat, chest infections, and conjunctivitis. Today, over-the-counter balms, rubs and ointments containing camphor are commonly used in Southeast Asia for medicinal purpose.

Camphor as an active ingredient in such externally used products is recognized by the US Food and Drug Administration (FDA) for its medicinal properties. Camphor reduces pain through distraction by activating the skin's temperature sensors and tricking the body into feeling cold. There is a maximum input that the nervous system can handle in one location, so forcing the body to focus on the coldness has the effect of masking the underlying pain.

## DRAGON'S BLOOD

Dragon's Blood is a deep red resin (Figure I.3) obtained mainly from the scales of the fruit from the rattan palm (a thorny vigorous climber in the rainforests) (Uphof, 1959), *Daemonorops draco*, found in Sumatra, Borneo and Peninsular Malaysia though resins were also occasionally obtained from some other rattan species as well.

Since ancient times, it has been used as an ethnomedicine (*sanguis draconis*) in China to heal wounds and stop bleeding (haemostasis), and for invigorating blood circulation for the treatment of traumatic injuries. Flavonoids and resin terpenoid acids are the main constituents of the resin and the flavonoids – dracorhodin (**3**), (2*S*)-5methoxyflavan-7-ol (**4**) and (2*S*)-5-methoxy-6-methylflavan-7-ol (**5**) are shown in Figure I.4. The resin also found its uses as varnish, incense and dye in the past. Dragon's Blood from *Daemonorops draco* has been found in shipwrecks in waters of Southeast Asia and has been identified by comparing the marker

**FIGURE I.3** Dragon's Blood resin (above) from *Daemonorpos draco* (Photo credit: Shuttlestock).

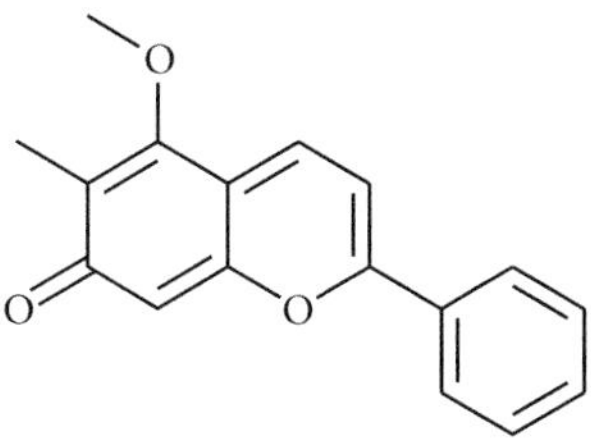

**FIGURE I.4** Some of the flavonoids – dracorhodin (**3**), (2*S*)-5methoxyflavan-7-ol (**4**) and (2*S*)-5-methoxy-6-methylflavan-7-ol (**5**) isolated from the resin of *Daemonorops draco*.

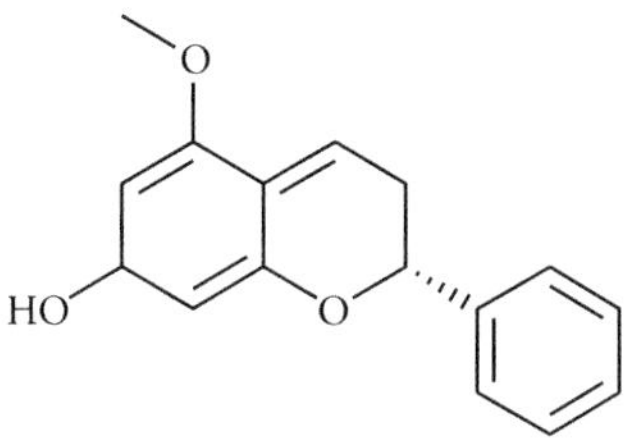

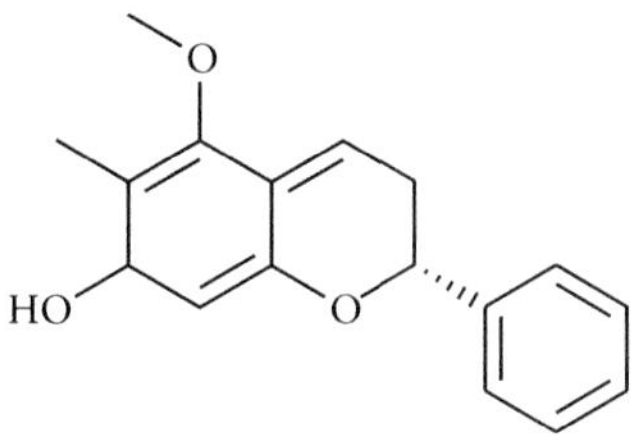

compounds in the resins from present-day materials as well as using nuclear magnetic resonance spectroscopy (Lambert *et al.*, 2016).
In present day, cosmetic products using Dragon's Blood from *Daemonorpos draco* have also been available in the market. However, the resin from *Daemonrpos draco* should not be confused with the similar bright red resins from other plant families often also referred to as Dragon's Blood, such as those from *Croton*, *Dracaena* and *Pterocarpus*. Few studies have been conducted on the medicinal properties of the *Daemonotops draco* resin but in a study done, it was shown to completely ameliorate pressure ulceration or bed sore in a patient with chronic pressure ulcer with tunneling in combination with local oxygen therapy, custom-built vacuum aspiration and anti-infection therapies (Ji *et al.*, 2015).

## MODERN USES OF PLANTS FROM BORNEO AS SOURCES OF MEDICINE

In modern times, plants from Borneo have been subject to modern screening techniques for pharmaceuticals. More recently, plants from Sarawak have been screened for anti-HIV (Currens *et al.*, 1996) and anti-cancer (Kim *et al.*, 2007) activities, culminating in the discoveries of anti-human immunodeficiency-virus type 1 (HIV Type 1) compounds – (+)-calanolide A (**6**) and (+)-calanolide B (**7**) from *Calophyllum lanigerum* var. *austrocoriaceum* and *Calophyllum teysmannii* var. *inophylloide* respectively (Currens *et al.*, 1996b) and the anti-cancer compound –

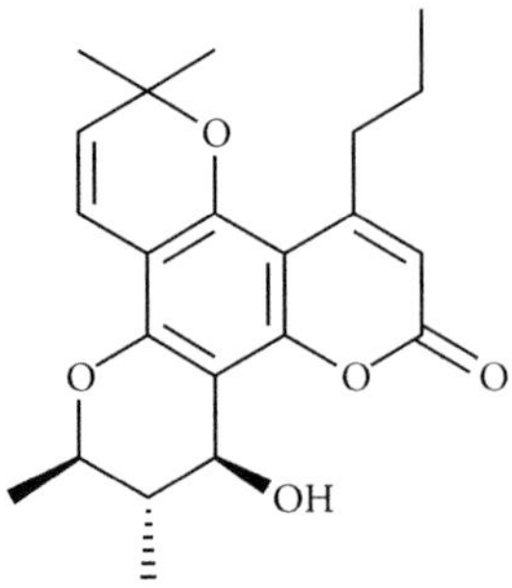

**FIGURE I.5** Anti-HIV Type 1 compounds – (+)-calanolide A (**6**) and (+)-calanolide B (**7**) isolated from the trees *Calophyllum lanigerum* var. *austrocoriaceum and Calophyllum teysmannii* var. i*nophylloide* respectively and the anti-cancer compound (protein synthesis translation inhibitor) silvestrol (**8**) isolated from the tree *Aglaia foveolata.*

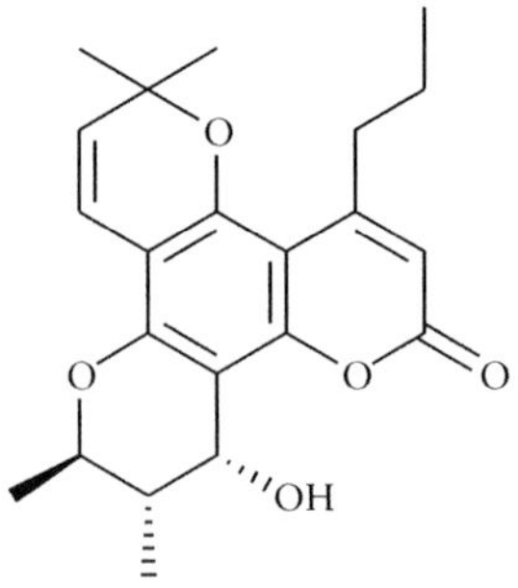

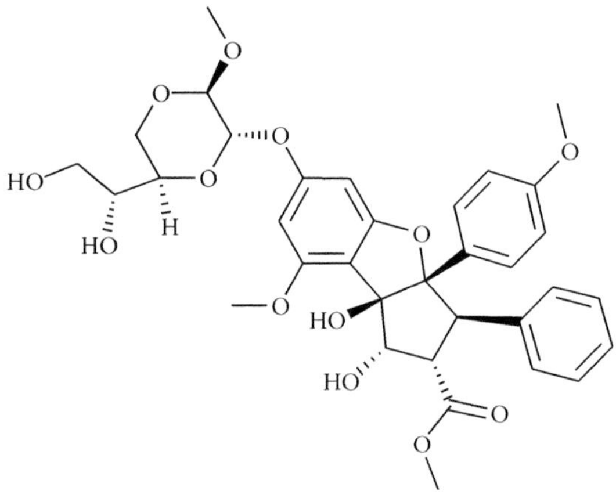

silvestrol from *Aglaia foveolata* (Kim *et al.*, 2007) as shown in Figure I.5. (±)-Calanolide A (**6**) and B (**7**) were also shown to display activity against *Mycobacterium tuberculosis* while silvestrol also showed *in vitro* activity against the Ebola (Biedenkopf *et al.*, 2016) and other viruses (Grünwelle, and Hartmann, 2017) by inhibiting their replication.

# 1 Anti-Malarial Plants

*Farid Kuswantoro and Stephen P. Teo*

## CONTENTS



## INTRODUCTION

Malaria has been a threat to mankind since ancient times. Even today, it is still a threat in developing tropical countries. The Centre for Disease Control (CDC) estimated that in 2016 alone there were 445,000 deaths caused by malaria worldwide. Malaria cases were also reported from Borneo. For instance, in 2013, a high number of malaria infections were reported in the Central and East Kalimantan provinces of Indonesian Borneo (Ompusunggu, 2015). Some plant species were utilized by people in Borneo to treat malaria (Leaman *et al.*, 1995). Below are described some of the anti-malarial plant species that have been used.

### *LANSIUM PARASITICUM* (OSBECK) K.C. SAHNI & BENNET

*Lansium parasiticum* (Osbeck) K.C. Sahni & Bennet (synonym: *Lansium domesticum*) is a fruit tree species (Figure 1.1) that belongs to the Meliaceae or Mahagony family (Anon., 2020).

**FIGURE 1.1** Tree (right) and leaves (left) of *Lansium parasiticum* (Photo credit: S. Teo).

## Traditional Use

Leaman *et al.* (1995) and Noorcahyati (2012) both reported the utilization of *L. parasiticum* as a remedy for the treatment malaria. In fact, the bark of this plant was used as a remedy for malaria throughout Borneo by the indigenous people before the advent of modern drugs such as chloroquine and others.

## Biological Activity

Various parts of *L. parasiticum* exhibited anti-plasmodial activity toward *Plasmodium falciparum*. The aqueous bark extract of *L. parasiticum* was reported by Leaman *et al.* (1995) and Omar *et al.* (2003) to display anti-plasmodial activity toward both chloroquine-resistant and sensitive strains of *P. falciparum*. Murnigsih *et al.* (2005), on the other hand, reported anti-plasmodial activity of *L. parasiticum* bark aqueous extract toward *P. falciparum*.

The leaves and fruit skin extract of *L. parasiticum* were also reported by Yapp and Yap (2003) to exhibit anti-plasmodial effects toward both chloroquine-resistant and the drug-sensitive *P. falciparum* strain, with the fruit skin extract also able to interrupt the parasite's life- cycle.

## Chemistry

Omar *et al.* (2003) reported that some compounds isolated from *L. parasiticum* extract were also shown to possess anti-plasmodial activity toward *Plasmodium falciparum* and *Plasmodium berghei*. Other research by Saewan *et al.* (2006) showed that some chemical compounds extracted from *L. parasiticum* seed also exhibited anti-plasmodial activity toward a multi-drug-resistant *P. falciparum* strain.

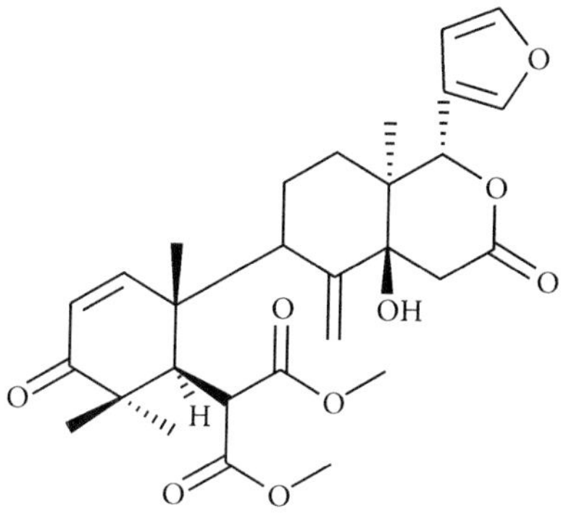

9

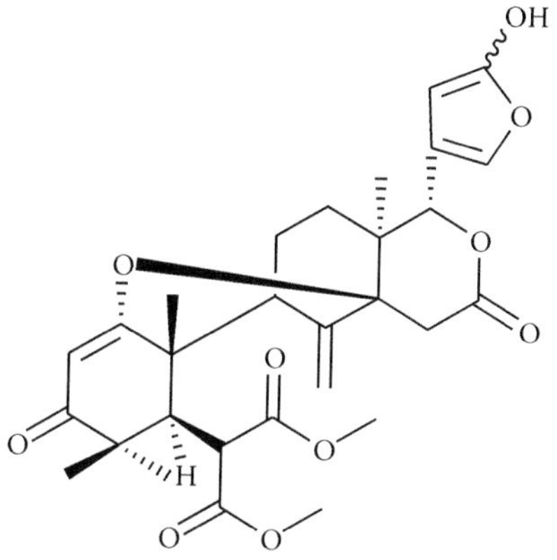

10

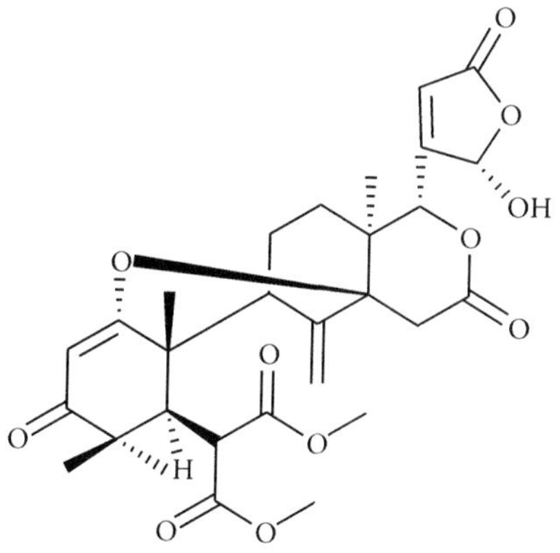

11

**FIGURE 1.2** Anti-malarial tetranortriterpenoids (limoniods) isolated from *Lansium parasiticum*–domesticulides (**9–11**).

Various types of triterpenoids such as cycloartanes, onoceranoids and tetranortriterpenoids (limonoids) including domesticulides B–D (**9–11**) (Figure 1.2) have been isolated from different parts of *L. parasiticum*. However, only tetranortriterpenoids have, so far, shown activity against the malarial parasite. Compounds such as Domesticulides B–D showed anti-malarial activity against *P. falciparum* with $IC_{50}$'s of 2.4–9.7 µg/mL (Saewan *et al.*, 2006).

## *Alstonia scholaris* (L.) R.Br.

*Alstonia scholaris* (L.) R.Br. is a tree species (Figure 1.3) belonging to the Apocynaceae family (Anon., 2020), Ahmad & Holdsworth (2003) and Noorcahyati (2012) reported traditional utilization of *Alstonia spp.* to cure various illnesses by people in Borneo.

**FIGURE 1.3** *Alstonia scholaris* (Photo credit: Shuttlestock).

## Traditional Use

Leaman *et al.* (1995), Ahmad & Holdsworth (2003) and Noorcahyati (2012) reported traditional uses of *A. scholaris* to cure malaria in Borneo.

## Biological Activity

Initial screenings with extracts of *A. scholaris* did not show any significant antimalarial activity. Leaman *et al.* (1995) reported little anti-plasmodial effect of *A. scholaris* leaves and bark toward both chloroquine-resistant and sensitive strains of *P. falciparum*. Furthermore, little antiplasmodial activity toward *P. falciparum* was also reported from a methanol extract of *A. scholaris* leaves, stem bark and root bark by Keawpradub *et al.* (1999). Gandhi &Vinayak (1990) and Pratyush *et al.* (2011) reported no anti-plasmodial activity was exhibited by the bark extracts of *A. scholaris* toward mice infected by *P. berghei* despite the methanol extract's ability to both improve the condition and delay death of the mice.

## Chemistry

Wright *et al.* (1993) noted that the major alkaloid present in various species of *Alstonia* was echitamine, but it displayed little anti-malarial activity. However, Hadi (2009) found two alkaloids, mataranines A (**12**) and B (**13**), (Figure 1.4) isolated from the leaves of a young *A. scholaris* tree, which showed activities

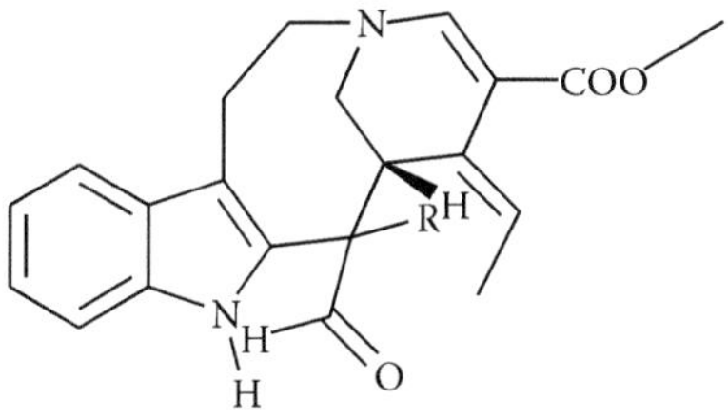

**12 Mataranine A, R = $H_\beta$**

**13 Mataranine B, R = $H_\alpha$**

**FIGURE 1.4** The two anti-malarial diastereomericindole alkaloids isolated from *A. scholaris*–mataranines A (**12**) and B (**13**).

against *P. falciparum* – KI strain with an $EC_{50}$ of 2.6 μg/mL (7.4 μM) and against TM4 strain with $EC_{50}$ value of 3.4 μg/mL (9.7 μM). It was noted that the anti-malarial compounds were only found in the young tree but not older tree (Hadi, 2009). Therefore the age of the plant is a factor in producing the anti-malarial alkaloids.

## *Eurycoma longifolia* Jack

*Eurycoma longifolia* Jack is a plant species that belongs to the Simaroubaceace family (Anon., 2020). In Borneo the species can be found throughout the island, especially in natural areas (Kartikawati *et al.*, 2014; Kuswantoro, 2017).

### Traditional Use

Setyowati *et al.* (2005), Setyowati (2010) and Noorcahyati (2012) all reported utilization of *E. longifolia* as a cure for malaria by people in Indonesian Borneo.

### Biological Activity

Numerous scientific publications suggested that *E. longifolia* has potential as a cure for malaria. Kardono *et al.* (1991), Kuo *et al.* (2003), Kuo *et al.* (2004), Chan *et al.* (1986) and Rehman *et al.* (2016) all reported anti-plasmodial activity of various chemical compounds isolated from the root extract of *E. longifolia* toward *P. falciparum.* Some quassinoids e.g. eurycomanone **(14),** 18 dehydro-6-α-hydroxyeurycomalactone **(15)** and eurycomanol **(16)** and alkaloids, e.g. β-carboline alkaloid **(17)** isolated from *E. longifolia* root extract (Figures 1.5 and 1.6) were also reported by Ang *et al.* (1995a) and Rehman *et al.* (2016); these compounds inhibited the growth of a Malaysian chloroquine-resistant strain of *P. falciparum.*

Ang *et al.* (1995b), Yusuf *et al.* (2013) and Rehman *et al.* (2016) reported the anti-plasmodial effects of *E. longifolia* root extract were dose dependent. Furthermore, Wernsdorfer *et al.* (2009) and Rehman *et al.* (2016) suggested that the synergism of the quassinoids or the presence of unidentified compounds was the reason for *E. longifolia* root extract having higher anti-plasmodial activity compared with the expected anti-plasmodial effect of the quassinoids from the

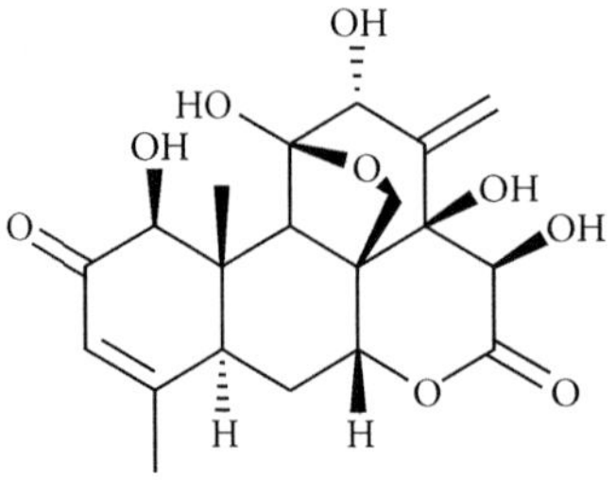

14

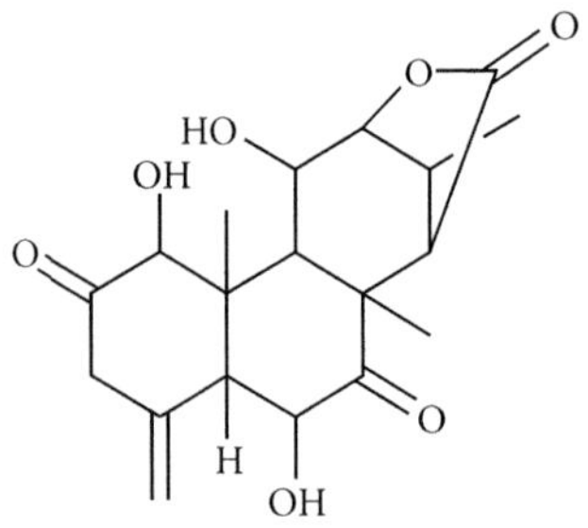

15

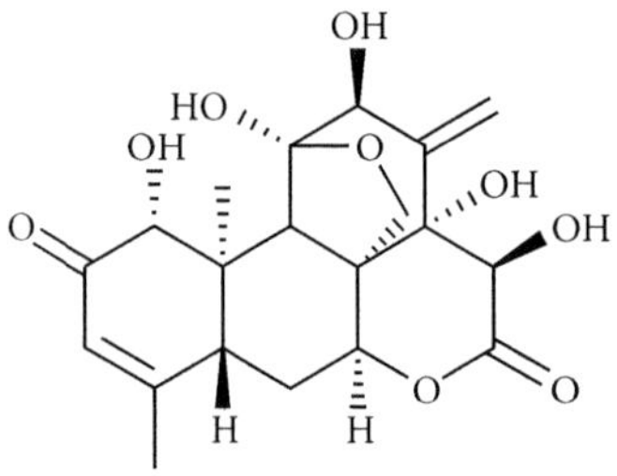

16

**FIGURE 1.5** Some examples of quassinoids isolated from *Eurycoma longifolia* with anti-malarial activities – eurycomanone (**14**), 18 dehydro-6-α-hydroxyeurycomalactone (15) and eurycomanol (**16**) (Kardano *et al.*, 1991).

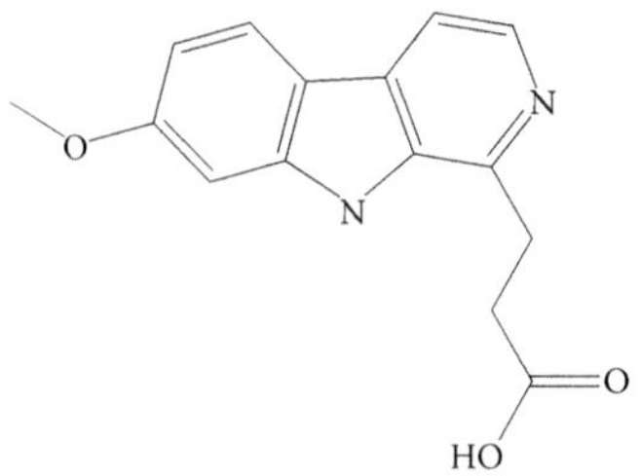

17

**FIGURE 1.6** A β-carboline alkaloid, 7-methoxy-beta-carboline-1-propionic acid (**17**), isolated from *Eurycoma longifolia* with anti-malarial activity (Kuo *et al.*, 2003).

respective extract. Other reports by Mohd. Ridzuan *et al.* (2005) and Rehman *et al.* (2016) stated that the anti-plasmodial effect of *E. longifolia* methanol extract toward *P. falciparum* might be caused by the extract's ability to affect glutathione content of both the parasite and the host.

However, anti-plasmodial activity was also reported to be present in other part of *E. longifolia*. Nguyen-Pouplin *et al.* (2007) reported an anti-plasmodial effect toward *P. falciparum* present in the extract of *E. longifolia* bark and leaves. Furthermore, Jiwajinda *et al.* (2002) and Rehman *et al.* (2016) also reported that quassinoids isolated from the leaves of *E. longifolia* exhibited antiplasmodial activity toward *P. falciparum.*

## *Tinospora crispa* (L.) Hook. f. & Thomson

*Tinospora crispa* (L.) Hook. f. & Thomson is a medicinal plant that belongs to Menispermaceae family (Figure 1.7; Anon., 2020).

### Traditional Use

Ahmad and Holdsworth (2003), Setyowati *et al.* (2005) and Falah *et al.* (2013) all reported *T. crispa* utilization to cure malaria by people of Borneo.

### Biological Activity

Bertani *et al.* (2005) and Ahmad *et al.* (2016) reported anti-plasmodial activity of *T. crispa* aqueous extract toward *P. falciparum*. The methanol extract of *T. crispa* was also reported by Rahman *et al.* (1999), Ihwan *et al.* (2014) and Ahmad *et al.* (2016) exhibited anti-plasmodial activity toward *P. falciparum*.

An anti-plasmodial effect toward *Plasmodium yoelii* was reported by Bertani *et al.* (2005), Rungruang and Boonmars (2009) and Ahmad *et al.* (2016) in the extract of *T. crispa*. Rahman *et al.* (1999), Niljan *et al.* (2014), and Ahmad *et al.*

**FIGURE 1.7** *Tinospora crispa* (stem) for sale at a market in central Sarawak.

(2016) also reported also an extract of the plant to display antiplasmodial activity toward *P. berghei.*

### Chemistry

There are no reports of compounds from *T. crispa* for anti-plasmodial activity.

## CARICA PAPAYA L.

*Carica papaya* L. is a popular cultivated fruit species throughout the tropics.

### Traditional Uses

An aqueous extract from the young leaves of *C. papaya* has been used to treat various diseases including malaria. Takoy *et al.* (2013) mentioned *C. papaya* as one of the plants used to treat malaria by people of Dayak Seberuang in West Kalimantan Province Indonesia.

### Chemistry

The alkaloid carpaine, which has also demonstrated anti-viral activity, is responsible for the anti-malarial activity (Tenga *et al.*, 2019).

Based on a study of a collection of anti-malarial plants from the Balikpapan Botanical Gardens, the majority of the plant parts used for the treatment of malaria is the bark and this is followed by the seeds and roots, whilst the least used is the leaves as shown in Figure 1.8. In terms of plant families, Apocynaceae and Lamiaceae contributed to more than one species for use against malaria and the rest were represented by one species each, i.e., Meliaceae, Rubiaceae, Simourabaceae, Arecaceae and Euphorbiaceae (Figure 1.8). Table 1.1 shows the *in situ* collection of anti-malarial plants used by the various ethnic groups in Kalimantan at the Balikpapan Botanical Gardens (Kuswantoro, 2017).

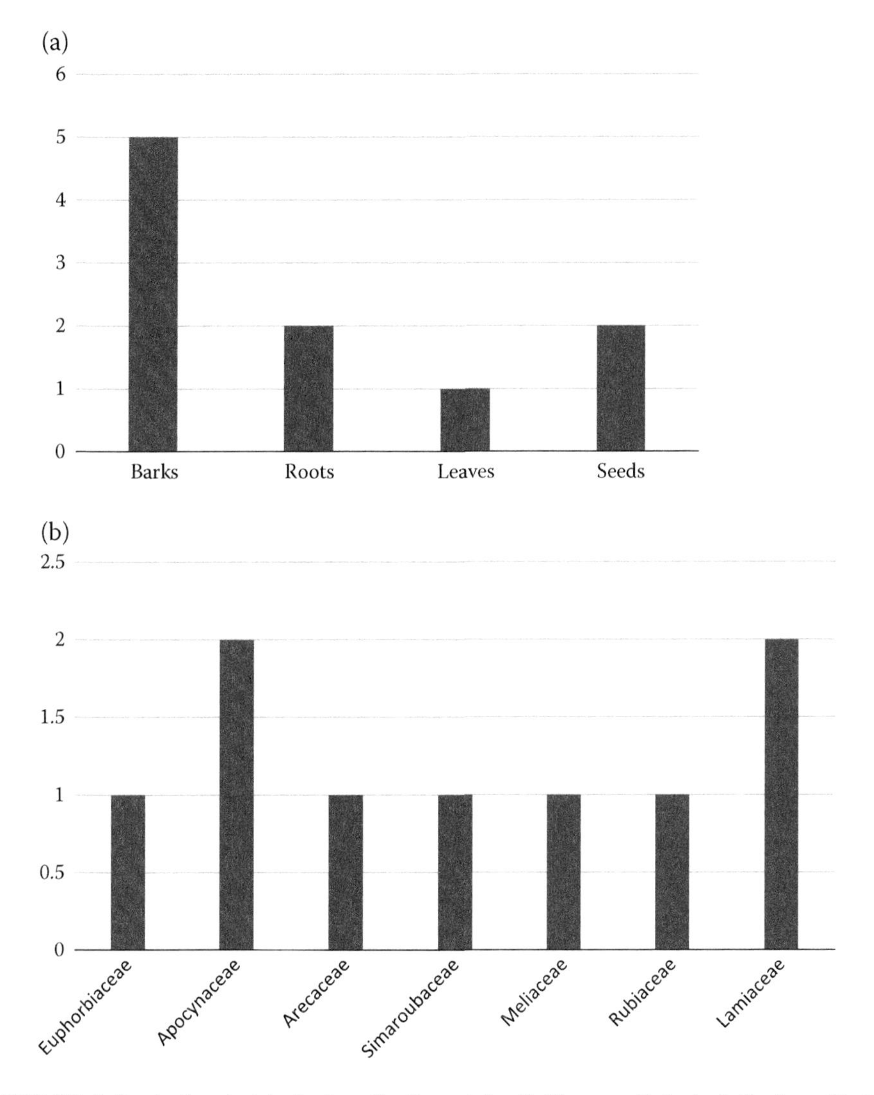


**FIGURE 1.8** Anti-malarial plants collection at the Balikpapan Botanical Gardens, East Kalimantan – plants parts used (A) and the families of the anti-malarial plants (B) (Kuswantoro, 2017).

**TABLE 1.1**
***In Situ* Collection of Anti-Malarial Plants Used by the Various Tribes in Borneo**

| Scientific Name | Local Name | Families | Used Parts | Used Procedure | Used by | References |
|---|---|---|---|---|---|---|
| *Aleurites moluccanus* (L.) Willd. | Perija, Kemiri | Euphorbiaceae | Barks | Boiled to drink | Banjar and Dayak ethnic | Noorcahyati (2012) |
| *Alstonia scholaris* L. R.Br. | Pulai | Apocynaceae | Barks | Boiled to drink | Kalimantan ethnics | Noorcahyati (2012) |
| *Alstonia angustiloba* Miq. | Pulai | Apocynaceae | Barks, Roots | Boiled to drink | Kalimantan ethnics | Noorcahyati (2012), Kustiawan (2007) |
| *Areca catechu* L. | Pinang | Arecaceae | Seeds | Boiled to drink | Dayak Tunjung | Setyowati (2010) |
| *Eurycoma longifolia* Jack | Pasak Bumi | Simaroubaceae | Roots | Boiled to drink | Kalimantan ethnics | Noorcahyati (2012), Setyowati (2010), Setyowati *et al.* (2005), Kustiawan (2007) |
| *Lansium parasiticum* (Osbeck) K.C. Sahni & Bennet | Langsat | Meliaceae | Barks | Boiled to drink | Dayak Seberuang and other Kalimantan ethnics | Takoy *et al.* (2013), Sari *et al.* (2014), Noorcahyati (2012) |
| *Morinda citrifolia* L. | Mengkudu | Rubiaceae | Leaves | Boiled to drink | Dayak Seberuang | Takoy *et al.* (2013) |
| *Peronema canescens* Jack | Sungkai | Lamiaceae | Barks | Boiled to drink | Kalimantan ethnics | Noorcahyati (2012) |
| *Vitex pinnata* L. | Alaban | Lamiaceae | Seeds | Mashed for eat | Kutai and Dayak ethnic | Noorcahyati (2012) |

*Note*: Collections were made from Balikpapan Botanical Gardens, Kalimantan, Indonesia (Kuswantoro, 2017).

# 2 Anti-Fungal Plants

*Stephen P. Teo*

## CONTENTS



## INTRODUCTION

Fungal infection of animals, including humans, known scientifically as mycoses are common and can be caused by a variety of environmental and physiological conditions. Colonization on a localized area on the skin as well as the inhalation of fungal spores may start persistent infections; therefore, mycoses often start in the lungs or on the skin.

In 2010, skin fungal infections ranked 4th amongst the most common diseases affecting approximately 984 million people (Hay *et al.*, 2010) while about 1.6 million people die each year of fungal infections (Anon., 2017). Fungal infection is also a leading cause of opportunistic infection for those with weakened immune system from AIDS, cancer and immune-compromised patients, amongst others. Cancer treatment and diabetes may also make a person more prone to fungal infections.

Various plant species are used in Borneo for treating different types of mycoses such as ring worm, jock itch, oral thrush and so forth.

**FIGURE 2.1** *Senna alata* (Photo credit: S. Teo).

## *Senna alata* L. (Roxb.)

*Senna alata* L. is a common wayside shrub (Figure 2.1) in Borneo, especially in open swampy areas. It is a naturalized species found introduced from tropical South America (Guyana, Brazil, Venezuela).

### Traditional Uses

The crushed leaves of *S. alata* are rubbed on the affected part the skin infected with ringworm fungus for a long time.

### Chemistry

Anthraquinones have been isolated from *S. alata* and have been shown to display antifungal activities (Hennebelle *et al.*, 1995; Wuthi-udomlert *et al.*, 2010). Anthraquinones isolated from the leaves of *S. alata* include aloe-emodin (**18**), rhein (**19**), chrysophanol (**20**) (Gritsanapan *et al.*, 1998; Fernand *et al.*, 2008), emodin (**21**) (Prasenjit *et al.*, 2016) and alatinone (**22**) (Hemlata,, 1993) as the main constituents (Figure 2.2).

### Biological Activities

An ethanolic leaf extract can inhibit dermatophytes (Wuthi-udomlert *et al.*, 2003; Sule *et al.*, 2010) and *Candida albicans* (Wuthi-udomlert *et al.*, 2003). Extracts from *S. alata* were shown to demonstrate activities against dermatophytes belonging to the genera *Trichophyton*, *Microsporum* and *Epidermophyton* which can lead to various types of dermatophytosis such as *Tinea capitis*, *Tinea cruris*, *Tinea corporis* and *Tinea pedis* that can affect different parts of the body. Sule *et al.* (2010) showed that the leaf exudates and the ethanol extract of the leaf of *S. alata* L. exhibited marked antifungal effects on *Microsporum canis*, *Trichophyton verrucosum*, *Trichophyton mentagrophytes* and *Epidermophyton verrucosum*. The ethanolic extract displayed the highest inhibition on *Trichophyton verrucosa* and *Epidermophyton verrucosum*

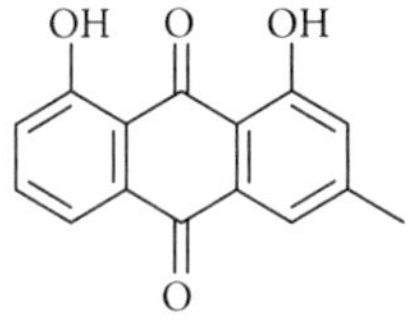

FIGURE 2.2 Various anthraquinones from *Senna alata* – aleo-emodin (**18**), rhein (**19**) chrysophanol (**20**), emodin (**21**) and alatinone (**22**).

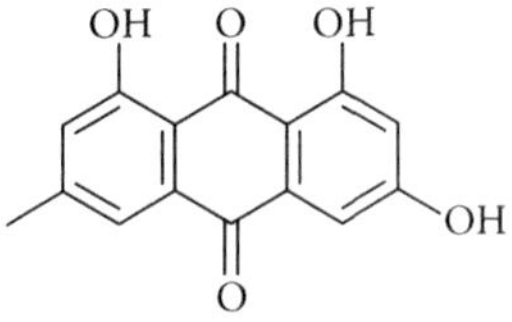

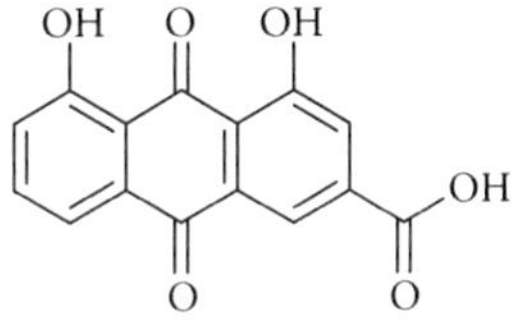

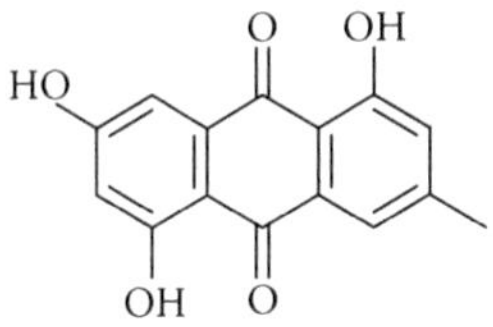

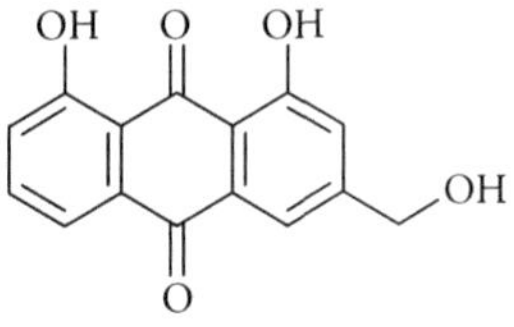

with a zone of inhibition of 20.5 and 20.0 mm, respectively while the Minimum Inhibitory Concentration (MIC) for all the tested dermatophytes were 5.0 m/mL (Sule *et al.*, 2010).

*S. alata* was among the 3 species of *Senna* screened by Phongpaichit *et al.* (2004) and was shown to be the most effective leaf extract against *Trichophyton rubrum and M. gypseum* with the 50% inhibition concentration ($IC_{50}$) of hyphal growth at 0.5 and 0.8 mg/mL, respectively. In addition, it was found that *S. alata* leaf extracts also affected *M. gypseum* conidial germination (Phongpaichit *et al.*, 2003). Microscopic observation revealed that the hyphae and macroconidia treated with leaf extracts shrank and collapsed, which was attributed to cell fluid leakage (Phongpaichit *et al.*, 2003).

On the other hand, Wuthi-udomlert *et al.* (2003) used various extraction methods, which all showed activities against 36 and 26 clinical isolates of dermatophytes and *C. albicans,* respectively.

## *Hymenocallis littoralis* (Jacq.) Salisb. and *Crinum asiaticum* L.

*Hymenocallis littoralis* (Jacq.) Salisb. and *Crinum asiaticum* L. (Figure 2.3) belong to the same family – Amaryllidaceae and are both tropical coastal beach species. *H. littoralis* occurs in the tropical part of South America while *C. asiaticum* is found in tropical Asia, the Pacific islands and Australia.

### Traditional Uses

Crushed plant was applied to affected areas for ring worm and jock itch.

### Chemistry

Phytochemicals reported from *H. littoralis* are mainly lycoriane alkaloids (Figure 2.4), namely littoraline (**23**), tazettine (**24**), pretazattine (**25**), macronine (**26**), lycorine (**27**),

**FIGURE 2.3** *Hymenocallis littoralis* (left) *and Crinum asiaticum* (right) (Photo credit: S. Teo).

homolycorhine **(28)**, lycorenine **(29)** and *O*-methyllycorenine **(30)** (Lin *et al.*, 1995). There were no studies that attributed the anti-*Candida* activities or other anti-fungal activities to the lycorrine alkaloids or other chemical compounds from this species.

### Biological Activities

*H. littoralis* displayed activity against the opportunistic yeast, *C. albicans* (Sundarasekar *et al.*, 2012). Crude methanolic extracts from the bulb, roots and leaves showed broad spectrum activities against Gram positive, Gram negative and the yeast, *C. albicans* with the best activities shown by the bulb methanolic extract against all groups. (Singh and Saxena, 2016). The methanolic bulb extract showed an inhibition zone of 20, 23, 19, 17 and 19 mm against *Staphylococcus aureus, Escherichia coli, Pseudomonas pneumonia, Salmonella typhimurium* and *C. albicans,* respectively (Singh and Saxena, 2016). Scanning electron microscopy revealed ruptures on the membrane in both *E. coli* and *C. albicans* (Rosli *et al.*, 2014). However, there was no report of screening of *H. littoralis* against pathogenic dermatophyte fungi and other fungi species.

A similar and related species in the same family, *C. asiaticum*, also showed good activities against *C. albicans* but few studies were performed to isolate the bioactive compounds and to screen for anti-fungal activities (Nair and van Staden, 2017).

## *Alpinia galanga* (L.) Wild.

*Alpinia galanga* (L.) Wild. or lengkuas is a member of the ginger family (Zingiberaceae). The rhizome of lengkuas is used for culinary purpose throughout Southeast Asia, including Borneo.

### Traditional Use

The rhizome of *A. galanga* (Figure 2.5) has been used against dermatophytes and rubbed against the skin of the affected parts throughout Borneo.

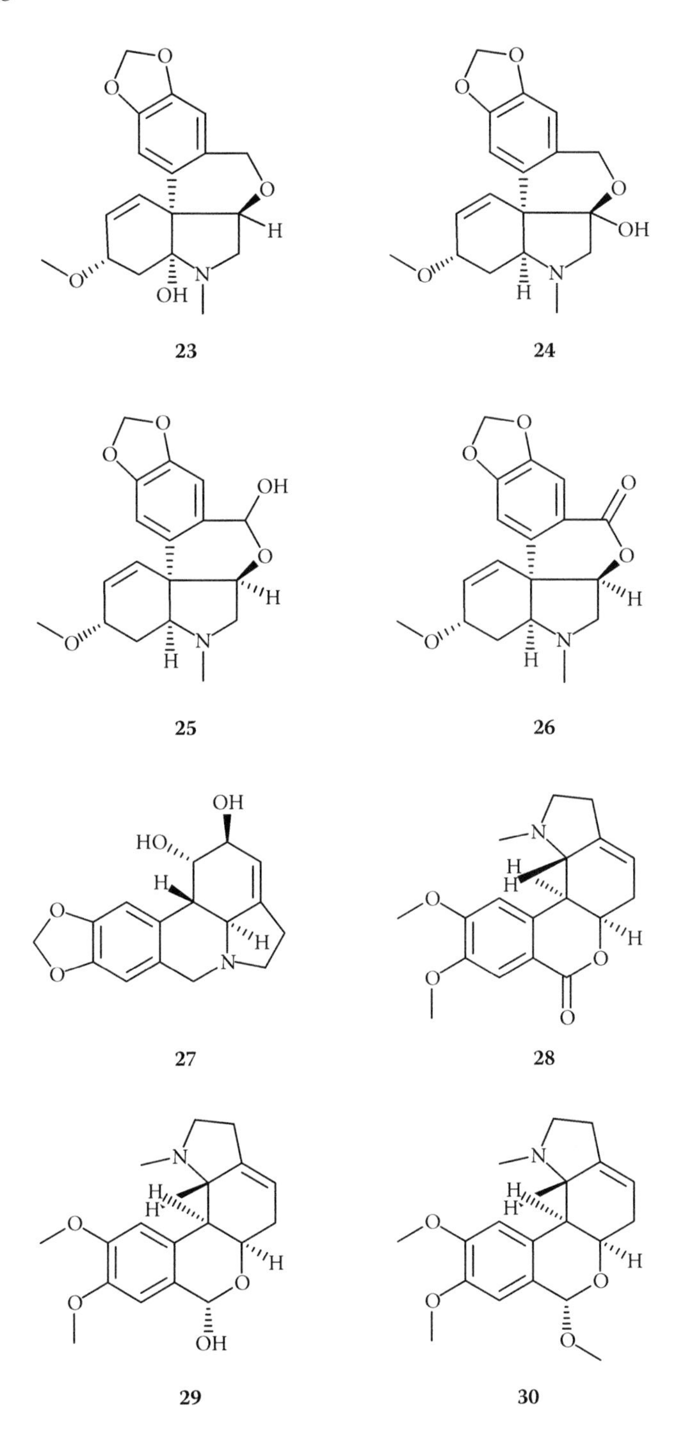


**FIGURE 2.4** Some examples of lycorine alkaloids from *Hymenocallis littoralis* - littoraline (**23**), tazettine (**24**), pretazattine (**25**), macronine (**26**), lycorine (**27**), homolycorhine (**28**), lycorenine (**29**) and *O*-methyllycorenine (**30**) (Lin *et al.*, 1995).

**FIGURE 2.5** Rhizomes of *Alpinia galanga* (Photo credit: S. Teo).

## Chemistry

The essential oil, acetoxychavicol acetate (**31**), isolated from the fresh and dried rhizhomes showed anti-fungal activity. Acetoxychavicol acetate (**31**) was active against the seven fungi tested and its MIC value for dermatophytes ranged from 50 to 250 μg/mL terpinen-4-ol (**32**) (Janssen and Scheffer).

The diterpene, (*E*)-8 beta,17-epoxylabd-12-ene-15,16-dial, isolated from *A. galanga* (Figure 2.6) displayed antifungal activity against the yeast, *C. albicans*, and this is due to a change of membrane permeability arising from membrane lipid alternation (Haraguchi *et al.*, 1995).

## Biological Activities

Khan *et al.* (2014) evaluated *A. galanga* extracts against Gram-positive and Gram-negative bacteria including multi-drug-resistant *S. aureus* (MRSA) and fungi such

**31** **32**

**FIGURE 2.6** Some of the anti-fungal compounds isolated from *Apinia galanga* – acetoxychavicol acetate (**31**) and terpinen-4-ol (**32**).

as *T. rubrum*, *M. canis* and *C. albicans*, using daptomycin, streptomycin and nystatin as standards, respectively. The hexane and methanol extracts showed significant antibacterial and antifungal activities with the zone of inhibition ranging from 14 to 26 mm (Khan *et al.*, 2014).

## *MUSA × PARADISIACA* L.

The cultivated banana species are either *Musa × paradisiaca* L., a hybrid between *Musa acuminata* and *Musa balbisiana* or *M. acuminata* alone (Figure 2.7). These are triploid plants and are cultivated throughout the tropics and subtropics of the world.

### Traditional Use

In Sarawak, the slimy exudates of cultivated banana stem pith were used to treat oral thrush in babies.

### Chemistry and Biological Activities

Anti-fungal activities in *Musa × paradisiaca* are often attributed to phytoalexins which are produced in response to fungal infection. In banana, the phytoalexins

**FIGURE 2.7** *Musa × paradisiaca* form typically found in Borneo (Photo credit: S. Teo).

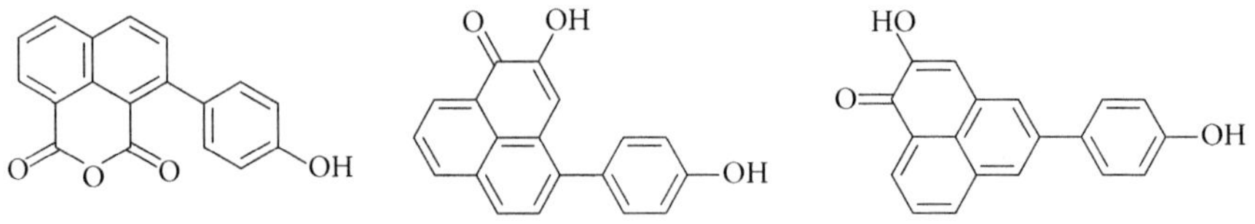

33 34 35

**FIGURE 2.8** Structures of phenalenones with anti-fungal activities from *Musa acuminata* – 2-(4′-hydroxyphenyl)-naphthalic anhydride (**33**) and the rhizomes of *Musa paradisiaca* – irenolone (**34**) and emenolone (**35**).

isolated are the phenalenone-type compounds (Figure 2.8) such as 2-(4′-hydroxyphenyl)-naphthalic anhydride (**33**) from the unripe green fruit of banana [*Musa acuminata* (AAA) cv Giant Cavendish] that shows resistance to infection by *Colletotrichum musae* (Hirai *et al.*, 1994) *and* irenolone (**34**) and emenolone (**35**) (Luis *et al.*, 1993).

# 3 Anti-Mycobacterial Plants

*Stephen P. Teo*

## CONTENTS



## INTRODUCTION

Infections caused by mycobacteria in humans were once considered dreaded diseases including tuberculosis (*Mycobacterium tuberculosis*) and leprosy (*Mycobacterium leprae*) until the arrival of better cures. Mycobacterial infections such as tuberculosis are still the leading cause of opportunistic infection amongst patients with weakened immune systems from infections such as HIV and immune-compromised patients undergoing, for example, chemotherapy.

### *HYDNOCARPUS WIGHTIANUS* BLUME

*Hydnocarpus wightianus* Blume or the chaulmoogra tree (Figure 3.1) is a tropical tree and endemic to India and the neighbouring regions such as Myanmar, but has been introduced to other tropical areas globally, mainly as a source of chaulmoogra oil for leprosy treatment in the days before the arrival of better drugs (Figure 3.2).

#### Traditional Uses

Chaulmoogra oil was used in Ayuverdic medicine as well as in traditional Chinese medicine for the treatment of leprosy and skin diseases. Chaulmoogra oil is found in different species of the genus *Hydnocarpus* where native species of *Hydnocarpus* also occur in Borneo, but the chief species from which the oil is obtained is the introduced species, *Hydnocarpus wrightiana*. In Borneo (Sarawak), chaulmoogra oil is extracted from the seeds of the fruits of *H. wrightiana* and was once a source of hope and treatment for lepers at the Rajah Charles Brooke Memorial Hospital in Kuching, Sarawak (Figure 3.3).

A single chaulmoogra tree introduced at the Rajah Brooke Memorial Hospital by migrants from China in the 1920s still exists to this day. Chaulmoogra oil was

**FIGURE 3.1** *Hydnocarpus wightianus* Blume or chaulmoogra tree (Photo credit: S. Teo).

**FIGURE 3.2** The fruit of *Hydnocarpus wrightiana* and the chaulmoogra oil is obtained from the seed of the fruits used in the treatment of leprosy (Photo credit: S. Teo).

**FIGURE 3.3** Rajah Brooke Memorial Hospital and the chaulmoogra tree in the background (arrow).

extracted and distilled from the seeds of the fruits (Figure 3.2) produced by the tree and in those days oil was administered as a treatment to cure leprosy.

The oil was used intramuscularly in the early part of the twentieth century against leprosy. The administration of the chaulmoogra oil was a painful procedure causing blistering underneath the skin. Only some patients were cured by the chaulmoogra oil treatment which could be due to a combination of the oil and the patients' own immune system. Patients came from other parts of Borneo (Brunei, Sabah and Kalimantan) as well as from neighbouring countries which formed one-eighth of the total patients treated at the leprosarium. (Jong, 2019). Instruments used to administer the oil are shown in Figure 3.4.

An African American, Alice Ball, later developed an ethyl ester of the oil in 1916 (Mendheim, 2007). Later in 1927, Burroughs Wellcome released the commercial preparation, sodium hydnocarpate marketed as Alepol, which produced lesser disagreeable symptoms of pain, swelling, irritating cough and blocking of the veins. There were reports of cure by doctors of leprosy in some patients after treatment with Alepol (Simpkin, 1928). Unfortunately, the water soluble form of the active principle in chaulmoogra oil was never introduced into Sarawak and Borneo.

The chaulmoogra oil treatment was later replaced by a new drug, dapsone, which was only partially effective due to the problem of drug resistance developed by *M. leprae.*

Times have changed and today leprosy is totally curable with the introduction of a multi-drug therapy made up of 3 antibiotics.

## Chemistry

The 3 major active compounds found in the chaulmoogra oil against *M. leprae* belong to the lipophilic cyclopentene-carboxylic acids i.e. hydnocarpic acid (**36**),

**FIGURE 3.4** Instruments used to administer chaulmoogra oil for the treatment of the lepers (Photo credit: S. Teo).

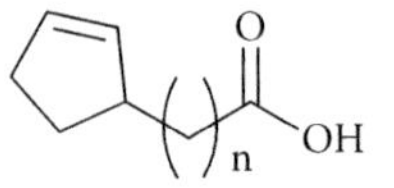

**36 n = 10, Hydnocarpic acid**

**37 n = 12, Chaulmoogric acid**

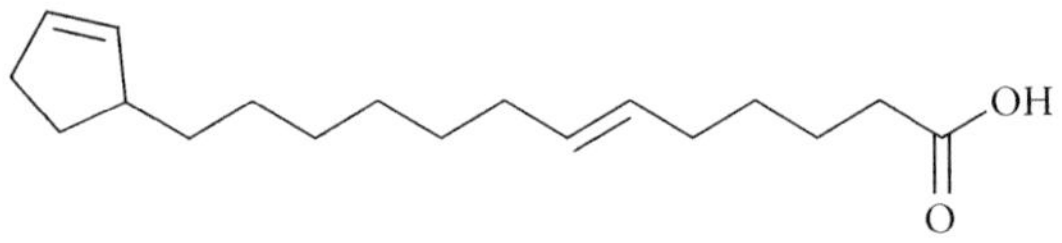

**Gorlic acid**

**38**

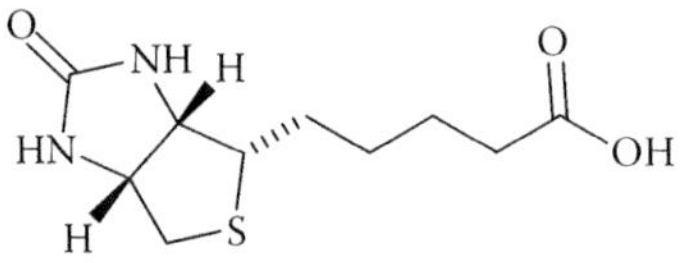

**39**

**FIGURE 3.5** Chemical structures of the 3 major active compounds found in the chaulmoogra oil – hydnocarpic acid (36), chaulmoogric acid (**37**) and gorlic acid (**38**) and structure of vitamin B7 or biotin (**39**).

chaulmoogric acid (**37**) and gorlic acid (**38**) (Figure 3.5) (Sengupta and Gupta, 1973). However, other analogues were described with different chain length or with a double bond at various distances from the carboxylic group (Badami and Paul, 1981).

### Biological Activities

Hydnocarpic acid (**36**) at a concentration of 30 μg/mL inhibited multiplication *in vitro* of 38 of 47 strains of 16 mycobacterial species (Levy *et al.*, 1973). Strains of *Mycobacterium gordonae* and *Mycobacterium intracellulare* were the most susceptible whist strains of *Mycobacterium kansasii* varied widely in susceptibility to hydnocarpic acid (**36**). Rapidly growing species tended to be more resistant than more slowly growing species (Levy *et al.*, 1973). Chaulmoogric acid (**37**) was slightly less active than hydnocarpic acid (**36**). Several straight chain fatty acids and dihydrochaulmoogric acid were found to be inactive (Levy *et al.*, 1973).

It was shown in laboratory studies that hydnocarpic acid (**36**) was able to inhibit the multiplication of a number of mycobacterial strains which were however not displayed by straight chain fatty acid or dihydrochaulmoogric acid (Jacobsen and Levy, 1973). Further experiments using *M. intracellulare* grown in Dubos medium supplemented with biotin (vitamin 7) (**39**) or without biotin revealed that hydnocarpic acid can inhibit those grown without biotin while the experiment supplemented with biotin (**39**) showed that biotin can interfere with the inhibitory action of hydnocarpic acid (Jacobsen and Levy, 1973). It was hypothesized that hydnocarpic acid (**36**) which shows some slight structural similarity to biotin is thought to interfere with either the coenzymatic reaction or the biosynthesis of biotin (**39**) in mycobacteria which is essential for their growth (Jacobsen and Levy, 1973).

Lipid analysis of *Mycobacterium vaccae*, grown in the presence of chaulmoogric acid (**37**), demonstrated that this cyclopentenyl fatty acid is taken up by the organism and incorporated into cellular phospholipids (membrane) and triacylglycerols (Cabot and Goucher, 1981). As cell growth is retarded by the addition of chaulmoogric acid (**37**) to the growth medium, it is possible that the antimicrobial properties of this compound result from a disruption of the membrane formation processes (Cabot and Goucher, 1981).

# 4 Anti-Viral Plants

*Stephen P. Teo*

## CONTENTS



## INTRODUCTION

Infectious diseases caused by viruses are serious threats to human. Among them are viral infections such as AIDS, dengue fever, zika, chikungunya, hepatitis, influenza, herpes, Ebola, and Covid-19.

Viral infection continues to be a pressing public health issue worldwide. For instance, the World Health Organization (2017a) noted that AIDS has, so far, caused more than 35 million deaths while there were about 36.7 million people infected with HIV at the end of 2016. The World Health Organization (2017b) data also showed that approximately 325 million people worldwide are infected with chronic hepatitis B virus (HBV) or hepatitis C virus (HCV).

### *CARICA PAPAYA* L.

*Carica papaya* L. (Figure 4.1) belongs to the family Caricaceae, is native to the northern part of South America including Mexico and was introduced to the tropical and subtropical countries including in Borneo (Malaysia, Brunei and Indonesia).

#### Traditional Uses

Leaves of *C. papaya* have been used traditionally for the treatment of dengue fever caused by the dengue virus, particularly in Sarawak.

**FIGURE 4.1** A *Carica papaya* tree with fruits (Photo credit: S. Teo).

## Biological Activity

An aqueous extract from papaya leaves was shown to increase the white blood cell and platelet count in dengue patients and with low cytotoxicity (Younas and Akhtar, 2014). The leaf extract also showed a significant inhibition of hemolysis *in vitro* and could prevent or minimize destabilization of biological membranes (Ranasinghe *et al.*, 2012). The British Medical Journal even provides a guideline for the use of papaya leaf extract for dengue fever patients (Hettige, 2015).

Flavonoids from papaya have been shown to inhibit the NS2B-NS3 protease in dengue 2 virus which is required for the viral replication and the flavonoid quercetin (**44**) was shown to have the highest binding energy with the protease (Padmanaban *et al.*, 2013). Molecular docking was done by the Universiti Putra Malaysia (UPM) using selected known compounds isolated from papaya (Radhakrishnan *et al.*, 2017). These included carpaine, dehydrocarpaine I and II, cardenolide, *p*-coumaric acid, chlorogenic acid, caricaxanthin, violaxanthin (**45**) and zeaxanthin against dengue, influenza and chikungunya viruses and demonstrated that violaxanthin was shown to have the highest interaction energy with that of dengue RNA dependent RNA polymerase (RdRp), influenza neuraminidase while caricaxanthin, violaxanthin and zeaxanthin exhibited interaction with Asn263 amino acid residue of chikungunya glycoprotein B chain.

However, the alkaloids isolated from papaya were never screened for activity against the dengue virus.

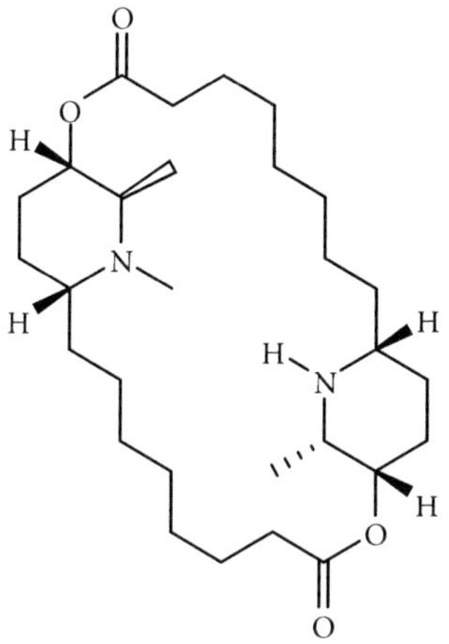

40

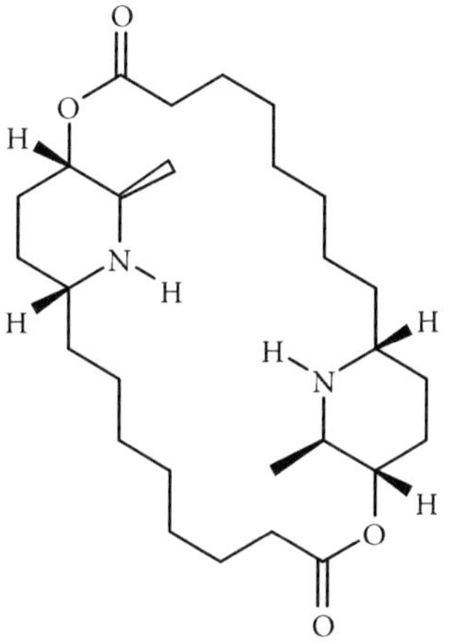

41

**FIGURE 4.2** Two isomers of the alkaloids, carpaine (**40**) and pseudocarpaine (**41**), isolated from the leaves of *Carica papaya.*

## Chemistry

Several compounds have been isolated from papaya which include alkaloids, flavonoids, steroid, coumarins and other phenolics. Alkaloids isolated from the leaves of papaya are carpaine (**40**) (Burdick, 1971; Xu *et al.*, 2004; Wang *et al.*, 2015), pseudocarpaine (**41**) (Govindachari *et al.*, 1954) (Figure 4.2) and dehydrocarpaine I and II (Tang, 1975). Among the flavonoids isolated were clitorin (**42**) and manghaslin (**43**) (Misnan *et al.*, 2015), quercetin 3(2G-rhamnosylrutinoside), kaempferol 3-(2G-rhamnosylrutinoside), quercetin 3-rutinoside, myricetin 3-rhamnoside, kaempferol 3-rutinoside, quercetin (**44**) and kaempferol (Nugroho *et al.*, 2017) whilst terpenoids isolated were caricaxanthin, violaxanthin (**45**) (Figure 4.3) zeaxanthin and danielone (Echeverri *et al.*, 1997). Other phenolics isolated were *p*-coumaric acid and chlorogenic acid.

# *Calophyllum lanigerum* Miq. var. *austrocoriaceum* P.F. Stevens and *Calophyllum teysmannii* Miq. var. *inophylloide* P.F. Stevens

Both species *Calophyllum lanigerum* Miq. and *C. teysmannii* Miq. are found in Peninsular Malaysia, and Borneo but their respective varieties (var. *austrocoriaceum* P.F. Stevens and var. *inophylloides* P.F. Stevens) are endemic to Borneo only (Figure 4.4).

## Traditional Uses

Both the species of *Calophyllum* were used traditionally as medicinal plants but were not used as an anti-viral The Kedayan in Sarawak used *Calophyllum soualattri* Burm. for antifungal infection of the skin (Chai, 2006).

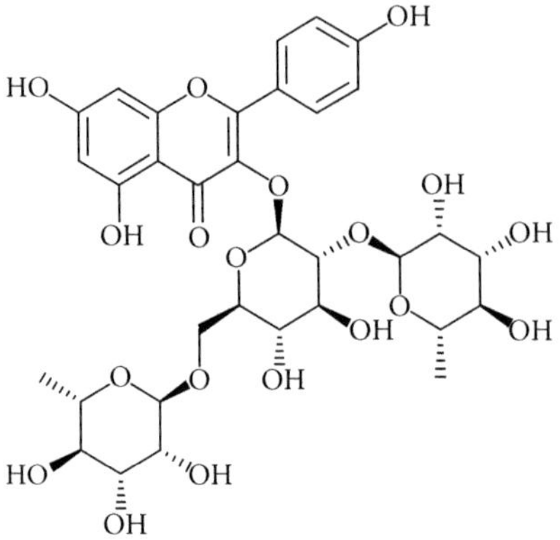

**FIGURE 4.3** Phenolics isolated from *Carica papaya*. Flavonoids – clitorin (**42**) manghaslin (**43**), quercetin (**44**) and terpenoid – violaxanthin (**45**).

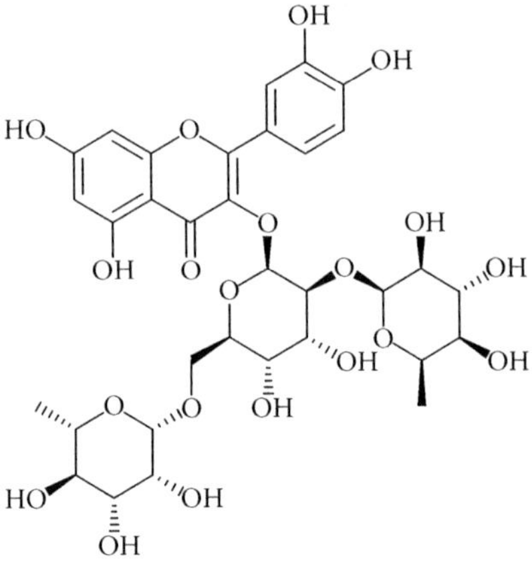

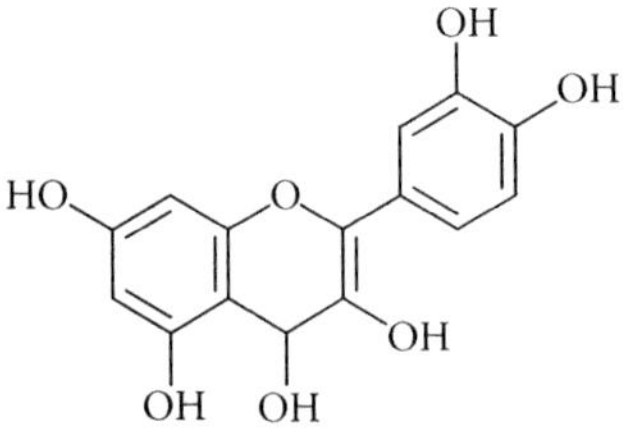

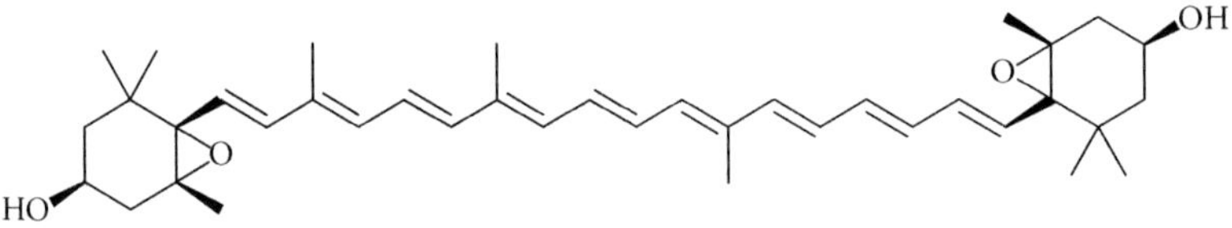

## Biological Activity

In 1986, a collaborative project between the Sarawak Forest Department, Malaysia and the National Cancer Institute of the United States was initiated with the aim of collecting and screening plants from the tropical rainforests of Sarawak for anti-cancer compounds. What was discovered instead were two species of *Calophyllum*, *C. lanigerum* and *C. teysmannii*, that harboured compounds which exhibited activities against the human-immunodeficeincy virus (HIV) type 1 (Mckee *et al.*, 1996). The bioactive compounds were (+)-calanolide A (**6**) and (+)-calanolide B (**7**) that displayed activities against HIV type 1 and are non-nucleoside inhibitors (NNRTIs) of the reverse transcriptase in the virus (Currens *et al.*, 1996b).

**FIGURE 4.4** Leaves of *Calophyllum teysmannii* var. *inophylloide* (above) and *Calophyllum lanigerum* var. *austrocoriaceum* (below) (Photo credit: S. Teo).

(+)-Calanolide A (**6**) showed synergistic anti-HIV activity when combined with nucleoside reverse transcriptase inhibitors, including AZT, ddI, ddC, and carbovir and also showed activity against strains of HIV which quickly became resistant to existing drugs (Buckheit *et al.*, 2001).

Apart from HIV Type 1, the calanolides (Figure 4.5) also displayed activity against *Mycobacterium tuberculosis* (mtb) (both susceptible and resistant strains). As mtb is the leading cause of opportunistic infection and a leading cause of mortality among HIV patients in the developing world, (+)-calanolide A (**6**) can serve two purposes.

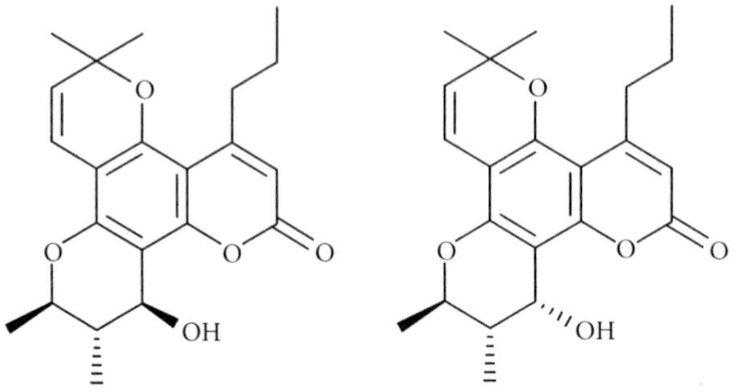

**FIGURE 4.5** Anti-HIV compounds – (+)-Calanolide A (**6**) and (+)-Calanolide B (**7**) isolated from the trees *Calophyllum lanigerum* var. *austrocoriaceum* P.F. Stevens and *Calophyllum teysmannii* var. *inophylloide* P.F. Stevens.

## Chemistry

Because (±)-calanolides (**6**) are rare in nature and present in small quantity, they were synthesized before further clinical work was carried out. The syntheses of (±)-calanolide A (**6**) and the related (±)-calanolides C and D were performed using a Lewis acid-promoted Claisen rearrangement in a short sequence to establish the chromene ring (Chenera *et al.*, 1993) and is summarised in Figure 4.6.

## Clinical Trials of (+)-calanolide A

Since (+)-calanolide A (**6**) gave the most potent activity, it was selected for clinical trials in the United States (Gallinis *et al.*, 1996). To further develop the (+)-calanolide A (**6**), a joint venture between Sarawak MediChem Pharmaceuticals, Inc. and MediChem Research, Inc., in Lemont, Illinois and the state of Sarawak, Malaysia was formed to advance its clinical development (GRAIN, 2017). In 1997, (+)-calanolide A (**6**) became the first compound from plants to reach clinical trials for HIV in the United States (Coleman, 1997). The viral life-cycle studies demonstrated that it acts at an early stage in the infection process, similar to the known existing HIV reverse transcriptase (RT) inhibitor. Results of a Phase 1A trial revealed that (+)-calanolide A (**6**) was well tolerated by the healthy volunteers. It was absorbed and circulated at a higher blood concentration than the data derived from animal studies. However, the most frequent side-effects were dizziness, oily aftertaste, headache, belching and nausea. Once or twice daily dosing was suggested based on pharmacokinetic studies (Creagh *et al.*, 2001). Phase II clinical trials on HIV-infected patients were conducted in both the United States and in Malaysia. It was suggested that (+)-calanolide A (**6**) may be used for combination therapy with existing drugs such as either the nucleoside AZT and/or protease inhibitors against HIV (Buckheit *et al.*, 2001). A further clinical trial was later conducted in Malaysia (Sarawak General Hospital, Kuching and Penang) and completed in 2016 using a derivative of the compound which displayed promising results.

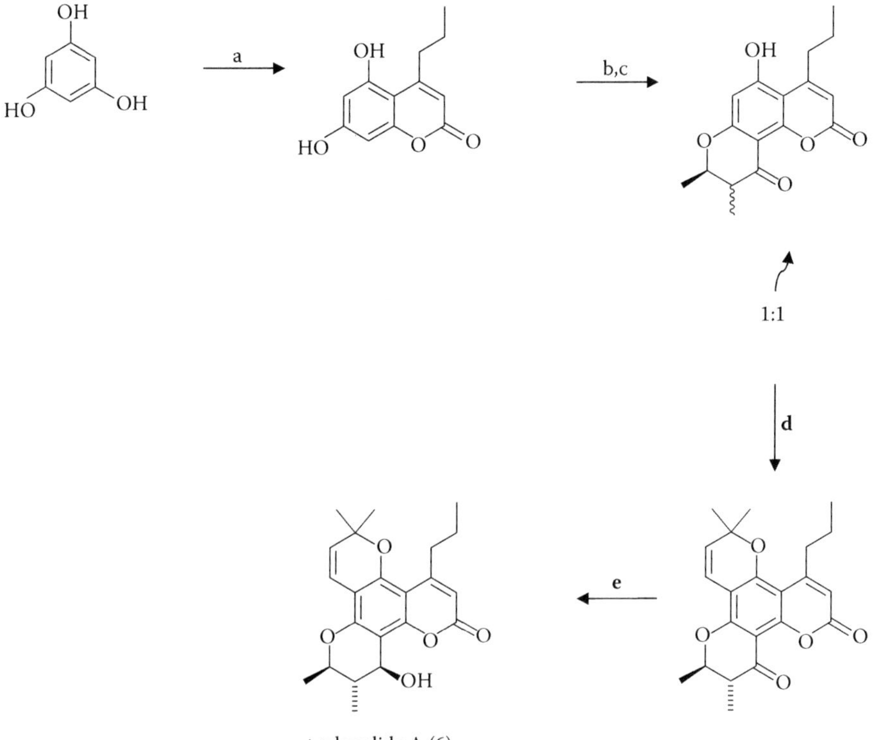

±calanolide A (6)

**FIGURE 4.6** Synthesis of (±) calanolide A (6) (Modified from Chenera *et al.*, 1993), (A) $C_3H_7COCH_2CO_2Et$, $CF_3SO_3H$ (neat), 0→25°C, 16 h (99%); (B) tigloyl chloride, $AlCl_3$ (4 equiv), $CS_2$ $PhNO_2$ 75°C, 14 h (87%); (C) $K_2CO_3$, 2-butanone, 70°C, 2 h (89%); (D) 3-chloro-3-methyl-1-butyne (5 equiv), $ZnCl_2$ (1.3 equiv), $K_2CO_3$ (2.5 equiv), $(n\text{-}Bu)_4NI$ (1 equiv), 2-butanone/DMF/$ET_2O$ (10:1:1), 70° C, 16 h (E) $NaBH_4$ (2 equiv), $CeCL_3(H_2O)_7$(1 equiv), EtOH, 25°C (59%).

## Other Antiviral Plants

In a screening of 50 herbal plants from the tropical rainforests of Sarawak, 11 crude extracts showed activities against the influenza virus $H_1N_1$ and $H_3N_3$ (Rajasekaran *et al.*, 2013). Assays for the 2 modes of viral inhibition (viral neuraminidase and haemagglutination assays) were carried out on the 11 extracts. All 11 extracts inhibited the enzymatic activity of viral neuraminidase while 4 extracts were shown to act through the haemagglutination inhibition (HI) pathway.

Silvestrol (**8**), a translation inhibitor, isolated from *Aglaia foveolata* from the Meliaceae plant family, which showed anti-cancer activity (Kim *et al.*, 2007) can also inhibit the replication of Ebola virus (Biedenkopf *et al.*, 2016) and other viruses. This will be discussed in the chapter for anti-cancer plants.

# 5 Anti-Cancer Plants

*Farid Kuswantoro and Stephen P. Teo*

## CONTENTS



## *DILLENIA SUFFRUTICOSA* (GRIFF.) MARTELLI

*Dillenia suffruticosa* (Griff.) Martelli (Figure 5.1) is a common wayside tree in Borneo belonging to the Dilleniaceae family (Anon., 2020).

### TRADITIONAL USES

*D. suffruticosa* was used by the Rungus people in Sabah to externally treat cancerous soreness (Ahmad and Holdsworth, 1995). The Jagoi Bidayuh in Sarawak use the pounded or chewed leaf as a poultice for cuts and wounds as well as for ringworm (Baling *et al.*, 2017).

### BIOLOGICAL ACTIVITY

In their review, Goh *et al.* (2017) covered some selected medicinal plant species of Brunei mentioned in the research of Yazan *et al.* (2015) which stated tumor reducing and metastasis inhibition properties of *D. suffruticosa* aqueous root extract toward 4t1 breast cancer in mice at 1000 mg/kg and 500 mg/kg, respectively. In the same review, Armania *et al.* (2013a) reported cytotoxicity properties

**FIGURE 5.1** *Dillenia suffruticosa* or simpoh (Photo credit: S. Teo).

of a methanol extract of *D. suffruticosa* root toward HeLa cancer cells while a sequential extract of the same part showed cytotoxicity toward various cancer cell lines such as the HeLa, MCF-7, MDA-MB231, A547 and HT29. Work by Armania *et al.* (2013b), also cited by Goh *et al.* (2017) mentioned that the D/F4 and EA/P2 fractions of *D. suffruticosa* root extract induced cell arrest toward MCF-7 estrogen positive breast cancer cells line whilst the D/F4 and D/F5 fractions of the same extract induced apoptosis toward the MDA-MB-231 estrogen negative breast cancer cells.

## Chemistry

Triterpenoids are common in *D. suffruticosa* and the few isolated so far included the well-known betulinic acid (**46**) (Farazimah *et al.*, 2018) (Figure 5.2).

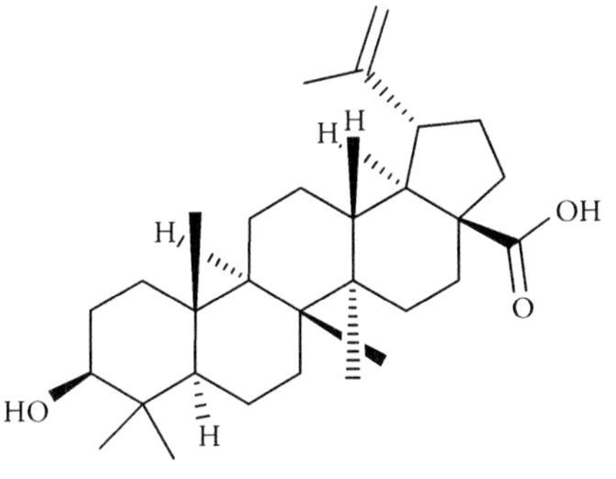

**FIGURE 5.2** Cytotoxic betulinic acid (**46**) from *Dillenia suffruticosa.*

## *ANNONA MURICATA* L.

*Annona muricata* L. (Figure 5.3) is a tropical fruit tree species belonging to the Annonaceae family (Anon., 2020).

### Traditional Uses

*A. muricata* was empirically used to treat cancer by people of Dayak Iban in West Kalimantan Province of Indonesia (Meliki *et al.*, 2013; Pradityo *et al.*, 2016).

**FIGURE 5.3** Fruits of *Annona muricata* (Photo credit: S. Teo).

### Biological Activity

An extensive number of scientific publications highlight the potential of *A. muricata* to treat cancer as reviewed by Rady *et al.* (2018) and Yajid *et al.* (2018). In their reviews, both groups compiled evidences of *A. muricata* anti-cancer properties toward various types of cancer such as breast, cervical, colorectal, lung, hepatic, pancreatic, prostate and melanoma cancers. *A. muricata* has been described as a safe anti-cancer agent as reported by Hansra *et al.* (2014) and cited by Rady *et al.* (2018), who reviewed its use in a woman with a chemotherapy-resistant breast cancer. However, safety of the materials must not be taken as guaranteed.

### Chemistry

In their reviews, Rady *et al.* (2018) and Yajid *et al.* (2018) also highlight annonaceous acetogenins (AGEs) as the major bioactive compounds of *A. muricata.* Some examples of acetogenins isolated from *A. muricata* from the fruit (Figure 5.3) are muricatetrocin A (**47**) (Chang and Wu, 2001), annoreticuin-9-one (**48**) (from seed) and sabadelin (**49**) (from flesh) (Ragasa *et al.*, 2012) (Figure 5.4).

## *AGLAIA FOVEOLATA* PANNELL

*Aglaia foveolata* Panell is a rainforest tree found throughout Borneo and belongs to the tree family Meliaceae.

### Traditional Uses

*A. foveolata* has traditionally been used by the Kelabits for stomach ailments but not as treatment for cancer.

### Chemistry and Biological Activity

Cencic *et al.* (2009) reported silvestrol (Figure 5.5) that been isolated from *A. foveolata* showed potent anti-cancer activity in human breast and prostate cancer xenograft models proved by increased apoptosis, decreased proliferation, and inhibition of angiogenesis. Further anti-cancer potential of silvestrol was reported by Kim *et al.* (2007) toward hormone-dependent human prostate cancer (LNCaP) cell lines.

## *EURYCOMA LONGIFOLIA* JACK

*Eurycoma longifolia* Jack (Figure 5.6) is a plant species belong to Simaroubaceae family (Anon., 2020).

### Traditional Uses

*E. longifolia* was used to treat many illnesses by people of Kalimantan (Noorcahyati, 2012).

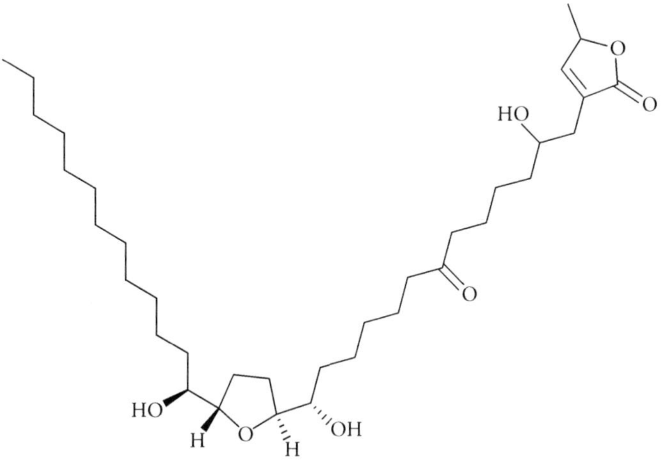

47

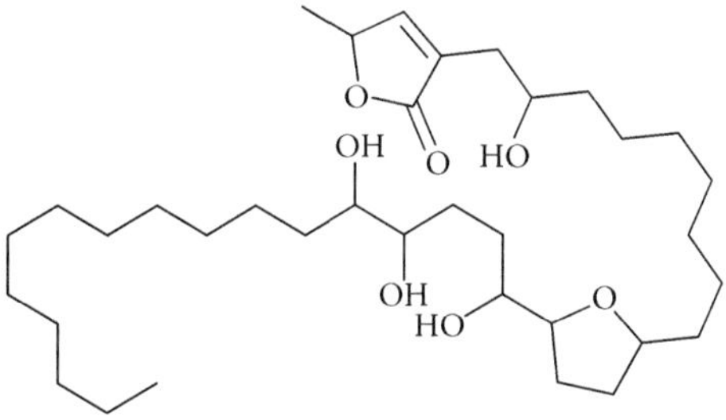

48

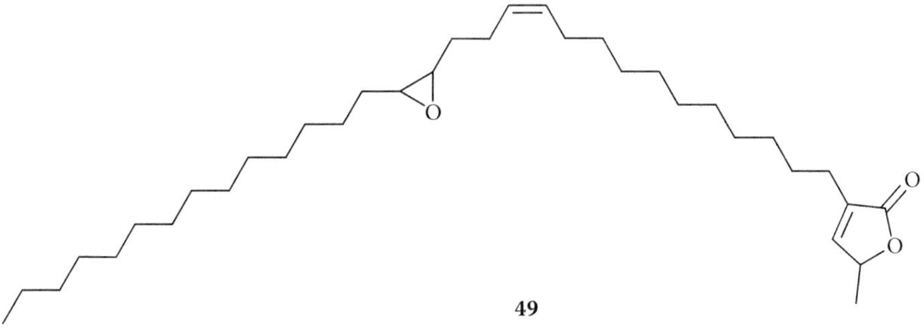

49

**FIGURE 5.4** Some examples of acetogenins isolated from *Annona muricata* – annoreticuin-9-one (**47**), muricatetrocin A (**48**) and sabadelin (**49**).

## Biological Activity

Extensive research has highlighted cytotoxicity activity of *E. longifolia*. In the review on pharmacological and phytochemical properties of *E. longifolia* Mohamed *et al.*

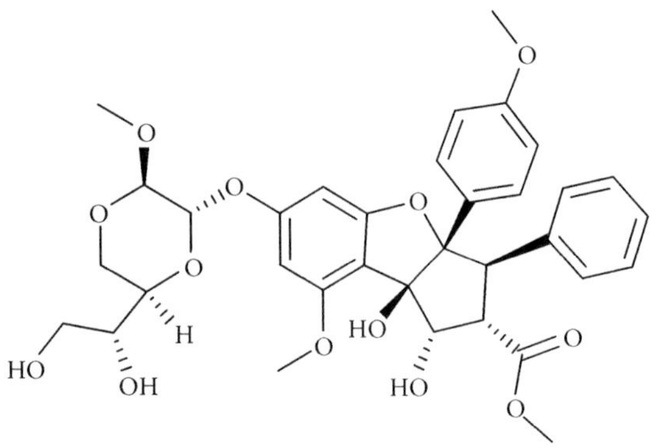

FIGURE 5.5 Structure of silvestrol (**8**).

FIGURE 5.6 *Eurycoma longifolia* (Photo credit: S. Teo).

(2015) mention research by Kuo *et al.* (2004) that reported cytotoxicity of *E. longifolia* compounds toward human lung and breast cancer cell lines. Moreover, review by Mohamed *et al.* (2015) also mentioned work of Nurhanan *et al.* (2005) which reported a cytotoxic effect of *E. longifolia* root extracts toward a number of cancer cell lines.

## Chemistry

The cytotoxic activities of *E. longifolia* are largely attributed to the quassinoids (Park *et al.*, 2014), which also displayed activities against malaria parasite; some of the structures of quassinoids are illustrated in Chapter 2.

# 6 Anti-Diabetic Plants

*Farid Kuswantoro and Stephen P. Teo*

## CONTENTS



## INTRODUCTION

In 2018, the World Health Organization (WHO) reported an increase in the number of people with diabetes from 108 million in 1980 to 422 million in 2014. Diabetes or *diabetes mellitus* is a metabolic disease due to the pancreas's inability to produce insulin or the ineffectiveness of insulin usage by the body (World Health Organization, 2018). Diabetes can be categorized into Type 1 diabetes caused by the inability of the pancreas to produce sufficient insulin with an unknown cause, Type 2 diabetes caused by the body invectiveness use of insulin and commonly the result of obesity and lack of exercise, and lastly *gestational diabetes* which can occur during pregnancy (World Health Organization, 2018). Some plant species of Borneo that have traditionally been used to manage diabetes are described below.

**FIGURE 6.1** *Alstonia angustifolia* (right) *Alstonia angustiloba* (left) (Photo credit: S. Teo and F. Kuswantoro).

## *ALSTONIA SPP.*

*Alstonia* is a genus of trees that belongs to the Apocynaceae family (Anon., 2020). There are at least 5 species of *Alstonia* in Borneo of which two are shown in Figure 6.1.

### Traditional Use

Ahmad & Holdsworth (1994), Ahmad & Holdsworth (2003), Noorcahyati (2012) and Zarta *et al.* (2018) reported that *A. scholaris, A. angustiloba* and *A. iwahigensis* were used by people of Borneo to treat diabetes.

### Biological Activity

Sinnathambi *et al.* (2010) and Meena *et al.* (2011) reported the ethanolic extract of *A. scholaris* leaves possess activity to lower blood glucose level in rats. Furthermore Jong-Anurakkun *et al.* (2007) and Meena *et al.* (2011) reported that compounds isolated from *A. scholaris* leaves extract showed inhibitory activity toward sucrase and maltase.

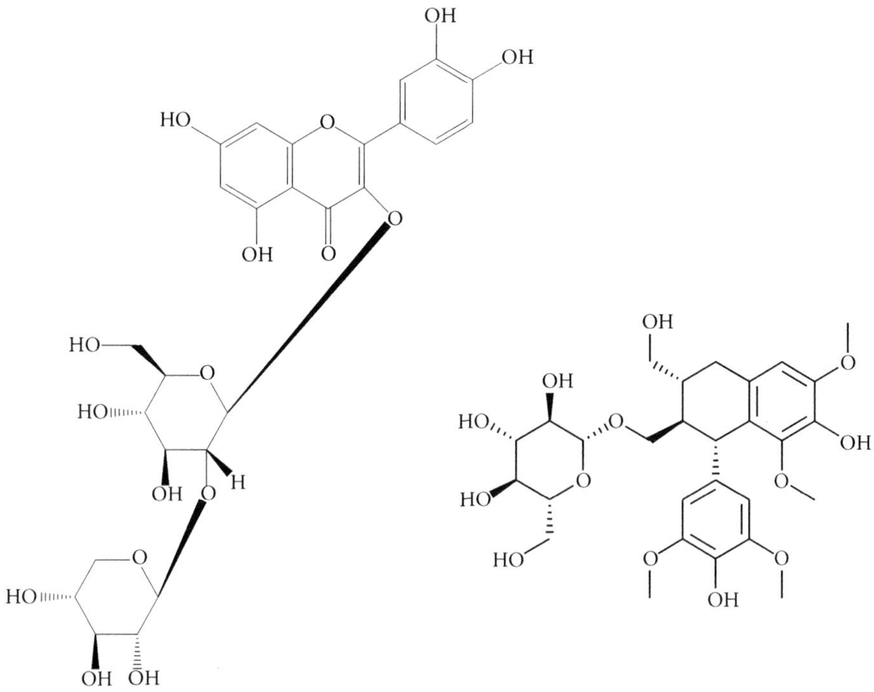

**FIGURE 6.2** Chemical structure of quercetin 3-*O*-β-D-xylopyranosyl (1‴ → 2″)-β-D-galactopyranoside (**50**) and (–)-lyoniresinol 3-*O*-β-D-glucopyranoside (**51**).

## Chemistry

Jong-Anurakkun *et al.* (2007) reported that quercetin 3-*O*-β-D-xylopyranosyl (1‴ → 2″)-β-D-galactopyranoside (**50**) (Figure 6.2) isolated from *A. scholaris* showed inhibitory activity toward maltase with $IC_{50}$ values of 1.96 mM, while another compound, namely (−)-lyoniresinol 3-*O*-β-D-glucopyranoside (**51**) exhibited an inhibitory activity against both sucrase and maltase with $IC_{50}$ values of 1.95 and 1.43 mM respectively.

## *Tinospora crispa* (L.) Hook. f. & Thomson

*Tinospora crispa* (L.) Hook. f. & Thomson is a medicinal plant that belongs to the Menispermaceae family (Anon., 2020). (Figure 6.3)

## Traditional Use

Noorcahyati (2012) reported the utilization of *T. crispa* by people of Kalimantan to treat diabetes. However, the use of *T. crispa* can probably cause hepatotoxicity due to clerodane furanoditerpenoids (Cachet *et al.*, 2018).

**FIGURE 6.3** *Tinospora crispa* (Photo credit: S. Teo).

## Biological Activity

Noor and Ashcroft (1989), Noor *et al.* (1989) and Noor and Ashcroft (1998) suggested that *in vivo* and *in vitro* antihyperglycemic activity of *T. crispa* stem extract was due to the extract ability to stimulate insulin release. Lam *et al.* (2012) reported increased glucose utilization in rats that treated on borapetoside C (**54**) isolated from *T. crispa*. Furthermore, Ruan *et al.* (2012) noted that other than being able to increase glucose utilization, borapetoside C (**54**) isolated from *T. crispa* was also able to increase the sensitivity of insulin as well as delayed insulin resistance.

## Chemistry

Hypoglycemic diterpenoids (Figure 6.4) have been isolated from *T. crispa*. Amongst them are borapetol B (**52**) Lokman *et al.* (2013), borapetoside A (**53**), borapetoside C (**54**) (Lam *et al.* 2012) and borapetoside E (**55**) (Xu *et al.*, 2017).

# *Artocarpus altilis* (Parkinson ex F.A. Zorn) Fosberg

*Artocarpus altilis* (Parkinson ex F.A. Zorn) Fosberg is a tree species belonging to the Moraceae family and is the accepted name of *A. communis* (Anon., 2020) (Figure 6.5).

## Traditional Use

Pradityo *et al.* (2016) and Noorcahyati (2012) reported traditional utilization of *A. altilis* leaves as medicine for diabetes by people of Kalimantan.

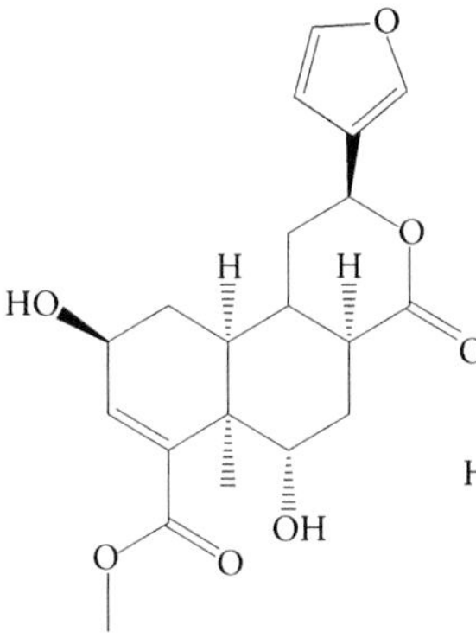

**FIGURE 6.4** Examples of diterpenoids isolated from *Tinospora crispa*, borapetol B (**52**), borapetoside A (**53**), C (54) and E (**55**).

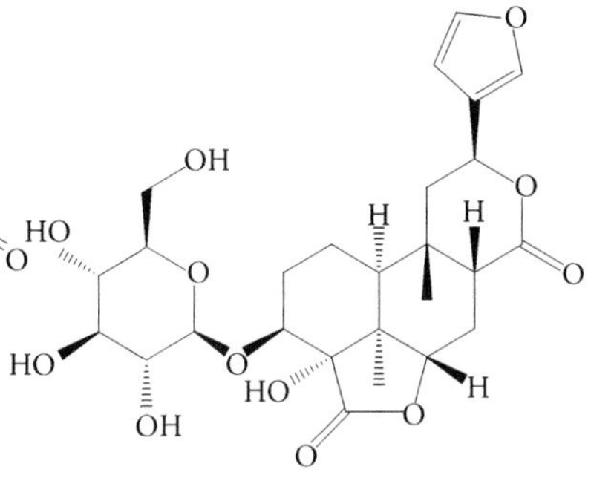

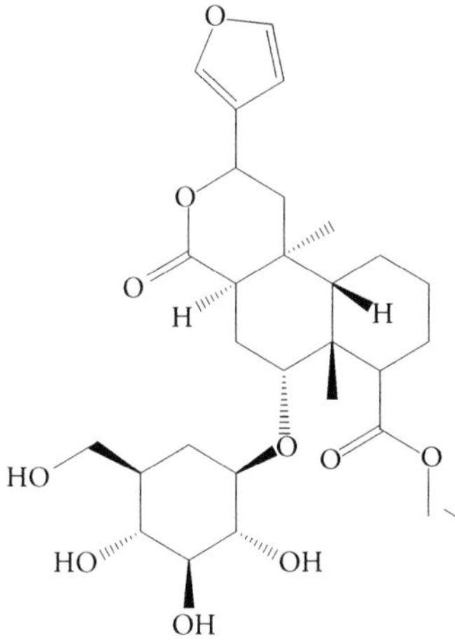

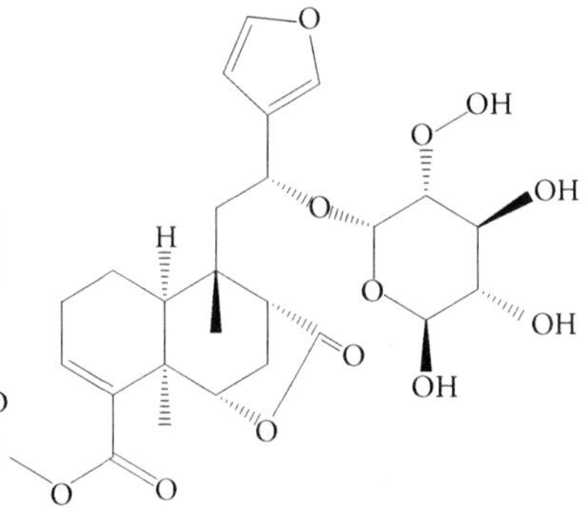

## Biological Activity

Nair *et al.* (2013) suggested potential anti-diabetic activity from methanolic extract of *A. altilis* fruit due to its inhibitory activity toward wheat alpha amylase and yeast alpha glucosidase.

## Chemistry

Flavonoids (geranyl aurones) had been isolated from *A. altilis* (Figure 6.6) (Huong *et al.*, 2012). The geranyl aurones H-I (**56-58**) played potent α-glucosidase inhibitory activity with $IC_{50}$ values ranging from 4.9 to 5.4 μM (Mai *et al.*, 2012).

**FIGURE 6.5** *Artocarpus altilis* (Photo credit: S. Teo)

## *Psidium guajava* L.

*Psidium guajava* L. (Figure 6.7), a guava tree with fruits is a tree introduced from South America and grown for its fruit. It is often grown as backyard fruit tree or as plantation fruit trees.

### Traditional Use

The water extract of the leaves of *P. guajava* has been used by the Kelabit in Sarawak to treat diabetes (Christensen, 2002).

### Biological Activity

Leaf water extract of *P. guajava* had been shown by various authors including Mukhtar *et al.* (2004) and Muthukumaran *et al.* (2018) to reduce hyperglycemia in alloxan-induced and STZ-induced diabetes respectively. A water extract of *P. guajava* leaves was screened for hypoglycemic activity on alloxan-induced diabetic rats. In both acute and sub-acute tests, the water extract, at an oral dose of 250 mg/kg, showed statistically significant hypoglycemic activity.

### Chemistry

The isolated bioactive compounds, lanosteroids (**59**–**62**), (Figure 6.8) have been found to show significant anti-diabetic activity. The selected bioactive

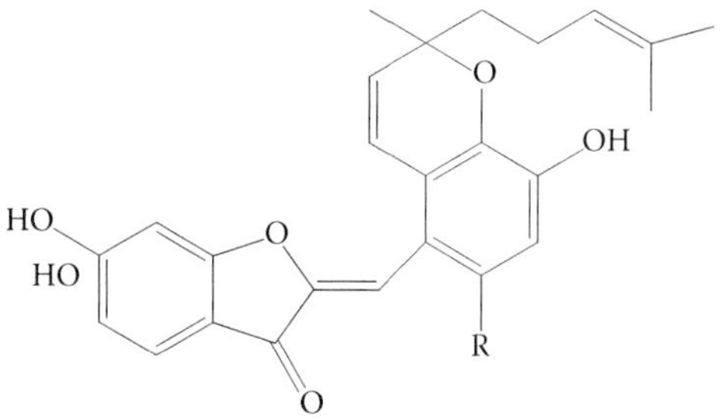

**56** R = H

**57** R = $OCH_3$

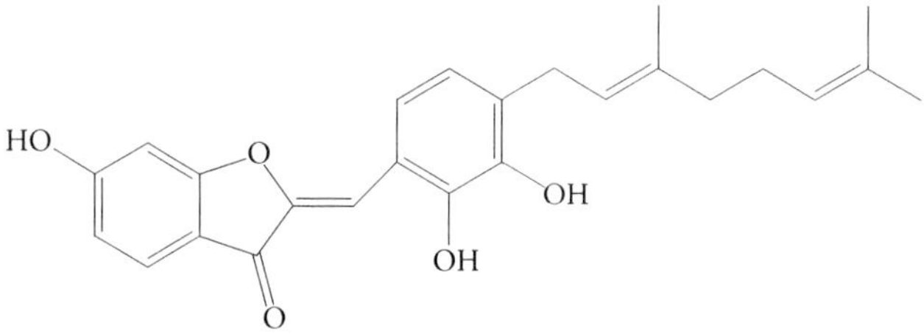

**58**

**FIGURE 6.6** Flavonoids isolated from *Artocarpus altilis* – prenylated aurone (artocarpaurone) - geranyl aurones altilisin H (**56**), I (**57**), and J (**58**), (Nguyen *et al.*, 2012).

**FIGURE 6.7** A guava tree with fruits (Photo credit: S. Teo).

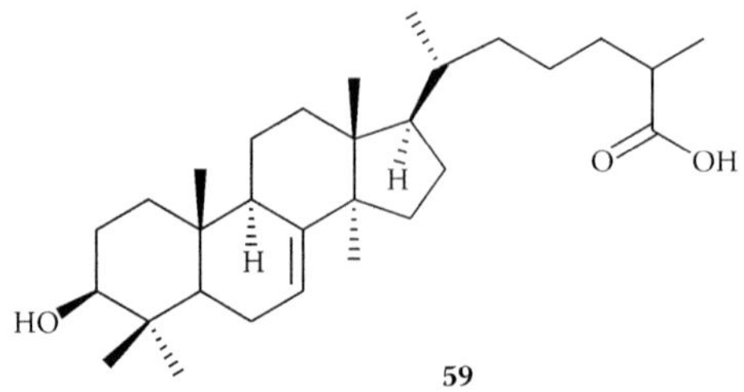

59

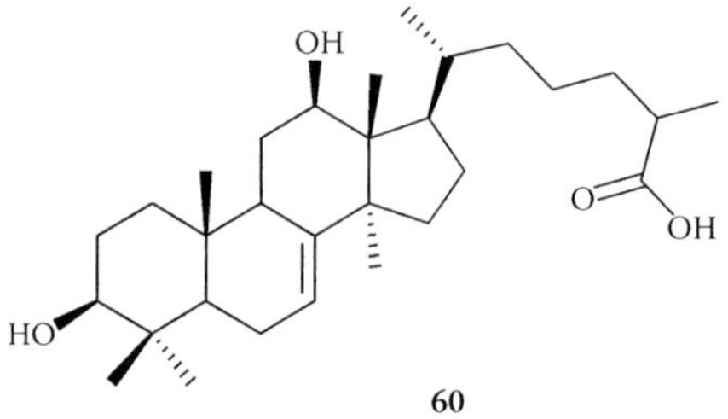

60

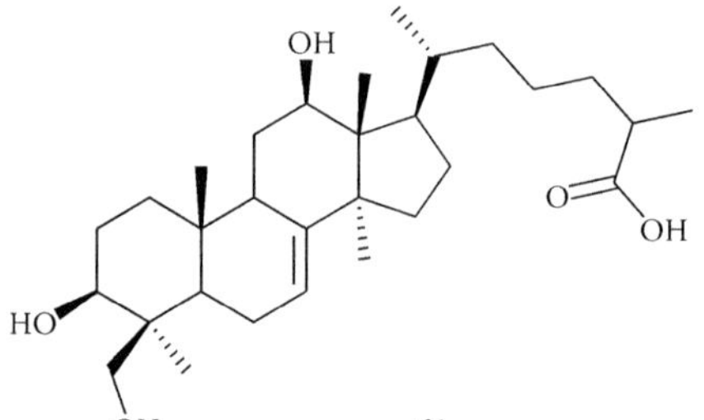

61

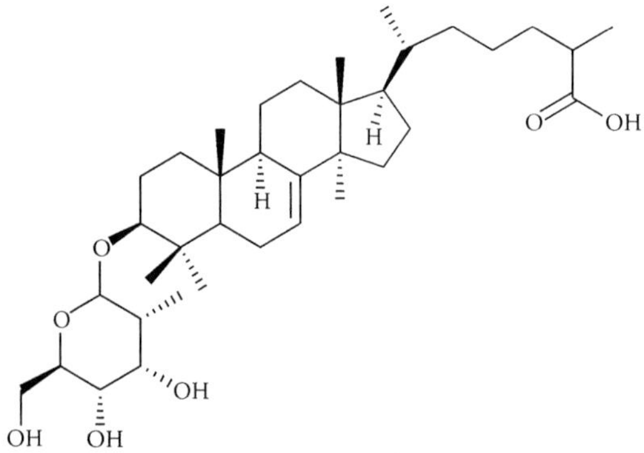

62

**FIGURE 6.8** Lanosteroids – lanost-7-en-3β-ol-26-oic acid (**59**), lanost-7-en-3β, 12β-diol-26-oic acid (**60**), lanost-7-en-3β, 12β, 29-triol-26-oic acid (**61**) and lanosteroid glycoside lanost-7-en-3β-ol-26-oic acid-3β-D-glucopyranoside (**62**).

compounds when orally administered, at a dose of 50 mg/kg for one week, significantly ($P < 0.001$) reduced the blood glucose levels in streptozotocin (STZ)-induced diabetic rats as compared with diabetic control rats, which showed that the bioactive compounds can control blood glucose level near to normal levels (Bagri *et al.*, 2016).

# 7 Plants for Fever Remedy

*Farid Kuswantoro*

## CONTENTS



## INTRODUCTION

Fever is one of the most common human medical symptoms. Fever can also occur as a result of various illness, from the common cold to more serious problem such as dengue and malaria, and thus, fever medicine is very dependent on the primary cause or illness. Fever can also cause an inflammatory reaction. Since it is very common for a human to become feverous, people of Borneo have used numerous plant species to traditionally relieve fever. In this chapter, we list some of them.

### *CENTELLA ASIATICA* (L.) URV.

*Centella asiatica* (L.) Urv. (Figure 7.1) is a member of Apiaceae family (Anon., 2020).

**FIGURE 7.1** *Centella asiatica* (Photo credit: S. Teo).

## Traditional Use

The indigenous Dayak Kanayatn and Dayak Ahe in West Kalimantan Province used *C. asiatica* to cure fever (Yusro *et al.*, 2014; Sari *et al.*, 2014). In Sabah and Sarawak it is widely used and consumed as salad.

## Biological Activity

A study by Wan *et al.* (2012) showed antipyretic activity of asiaticoside (**63**) extracted from *C. asiatica*.

## Chemistry

Asiaticoside (**63**) (Figure 7.2) is a triterpenoid with antipyretic activity in *C. asiatica* (Wan *et al.*, 2012).

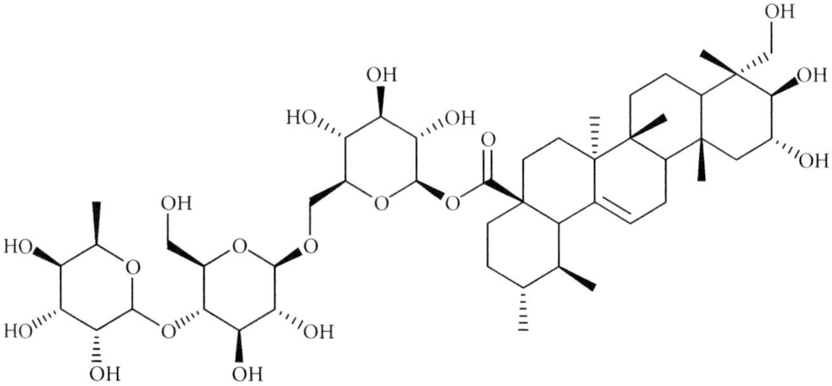

**63**

**FIGURE 7.2** The anti-pyretic triterpenoid, asiaticoside (**63**).

**FIGURE 7.3** *Kaempferia galanga* (Photo credit: K. Meekiong).

## *Kaempferia galanga* L.

*Kaempferia galanga* or Galangal (Figure 7.3) is a member of Zingiberaceae family (Anon., 2020) and is widely used for culinary purposes.

### Traditional Use

Dayak Kenayatn in West Kalimantan Province used Galangal rhizome to cure fever (Yusro *et al.*, 2014). Dayak Tomun of Central Kalimantan Province also used Galangal as a remedy for fever in children (Santoso *et al.*, 2019).

### Biological Activity

Anti-inflammatory activity from K. *galanga was exhibited by leaf aqueous extract* (Sulaiman *et al.*, 2008). A study done by Umar *et al.* (2012) reported anti-inflammatory activity of a *K. galanga* extract.

### Chemistry

Anti-inflammatory activity was due to the effect of ethyl-*p*-methoxycinnamate (**64**) (Figure 7.4) extracted from *K. galanga* rhizome as reported by Umar *et al.* (2012).

## *Hibiscus rosa-sinensis* L.

*Hibiscus rosa-sinensis* L. (Figure 7.5) belongs to the Malvaceae family (Anon., 2020).

**FIGURE 7.4** Structure of ethyl-*p*-methoxycinnamate (**64**).

**FIGURE 7.5** *Hibiscus rosa-sinensis* (Photo credit: S. Teo).

## Traditional Use

Several Dayak communities in West Kalimantan Province use *H. rosa-sinensis* leaf to cure fever (Diba *et al.*, 2013). *H. rosa-sinensis* flower was used by the Dayak Iban of the same province to treat fever-stricken children (Pradityo *et al.*, 2016). *H. rosa-sinensis* is also used as fever medicine by the Malay ethnic groups from the village of Serambai, West Kalimantan province (Sari *et al.*, 2014).

## Biological Activity

A significant antipyretic activity was attributed to aqueous extracts of *H. rosa-sinensis* root and leaf (Soni and Gupta, 2011; Daud *et al.*, 2016; Al-Snafi 2018). Low toxicity and high antipyretic activity of ethanolic extract of *H. rosa-sinensis* flower was also reported (Birari *et al.*, 2009; Al-Snafi 2018). The anti-inflammatory effect was also reported from an extract of *H. rosa-sinensis* root and flowers (Begum *et al.*, 2018; Birari *et al.*, 2009; Al-Snafi 2018).

### Chemistry

So far no compound with anti-pyretic and anti-inflammatory activity of *Hibiscus rosa-sinensis* has been isolated. However, Patel and Adhav (2016) suggested that glycosides, flavonoids, phytosterol, terpenoids, phenol and tannins found in this plant species correlate with its healing properties.

## *Blumea balsamifera* (L.) DC.

*Blumea balsamifera* (L.) DC. (Figure 7.6) belongs to the Compositae family (Anon., 2020).

### Traditional Use

Burnt leaf of *B. balsamifera* was used to treat fever by the people of Dayak Desa (Supiandi *et al.*, 2019). The Kadazan people of Sabah treated fever by consuming a concoction of *B. balsamifera* leaf as well as utilized fumes generated from a heated mixture of *B. balsamifera* (Ahmad and Holdsworth, 2003).

**FIGURE 7.6** *Blumea balsamifera* (Photo credit: S. Teo).

### Biological Activity

To the best of our knowledge, no antipyretic activity was recorded from *B. balsamifera*. However, in-vivo antipyretic activity toward Brewer's yeast-induced pyrexia was attributed to *B. densiflora* ethanol extract (Hossain *et al.*, 2016). The anti-inflammatory effect of *B. balsamifera* is due to the hexane leaf extract (Bantol *et al., 2017).*

### Chemistry

The study by Wang and Yu (2018) suggested that *B. balsamifera* essential oil from leaves contains caryophyllene, xanthoxylin, γ-eudesmol and α-cubenene with high antioxidant activity. The antioxidants are useful to maintain health and thus prevent fever.

## CYMBOPOGON CITRATUS (DC.) STAPF

*Cymbopogon citratus* (DC.) Stapf (Figure 7.7) or lemongrass or fevergrass is a plant species that belongs to the grass family of Poaceae (Anon., 2020).

### Traditional Use

The Murut people in Sabah drank a concoction of *C. citratus* to relieve fever (Ahmad and Holdsworth, 1994). Meanwhile, vapour from the heated mixture of *C. citratus* with *B. balsamifera* and some other plants species were used by the Kadazan people in Sabah as a fever remedy (Ahmad and Holdsworth, 2003).

### Biological Activity

Tarkang *et al.* (2015) showed that *C. citratus* exhibited the best antipyretic activity toward both D-amphetamine and Brewer's yeast-induced mice compared to the

**FIGURE 7.7** *Cymbopogon citratus*. (Photo credit: S. Teo).

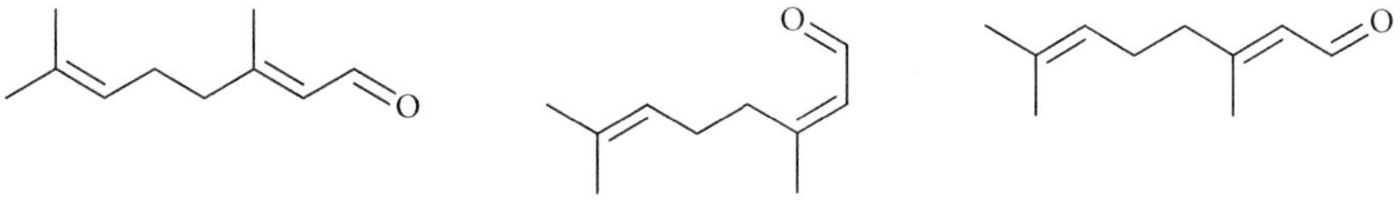

65 66 67

**FIGURE 7.8** Geranial (**65**) neral (**66**) and citral (**67**).

other tested plant species. *C. citratus* essential oil also exhibited anti-inflammatory activity (Gbenou *et al.*, 2013).

## Chemistry

Geranial (**65**), neral (**66**) and myrcene (**67**) (Figure 7.8) are the main compounds found in *C. citratus* essential oil (Gbenou *et al.*, 2013).

# 8 Anti-Hypertensive Plants

*Stephen P. Teo*

## CONTENTS



## INTRODUCTION

Hypertension is one of the most common cardiovascular diseases globally. It is often referred to as high blood pressure. It is a long-term condition whereby the blood pressure is elevated. However, in some cases, symptoms may not be

**FIGURE 8.1** *Andrographis paniculata*. (Photo credit: S. Teo).

displayed which can be a great risk and can also cause detrimental effects on the body in the long term.

In the region, several folklore medicines are recognized as useful for treating such diseases and are reviewed below.

## *Andrographis paniculata* (Burm. *f.*) Nees

*Andrographis paniculata* (Burm. *f.*) Nees (Figure 8.1) is native to South India and Sri Lanka but has since been introduced elsewhere, particularly in the tropics including Borneo.

### Traditional Uses

Leaves of *A. paniculata* are placed in hot water to make a tincture to treat hypertension (Christensen, 2002; Chai, 2006).

### Biological Activity

Ethanolic (90%) extract of *A. paniculata* exhibited a potent anti-hypertensive activity in phenylephrine-induced hypertensive Wistar rats (WKY) by decreasing systolic and diastolic blood pressures by up to 120% and 150%, respectively (Zhang and Tan, 1996). The optimum hypotensive dose determined was repeated in a study performed in spontaneously hypertensive rats (SHR) and their normotensive controls, Wistar Kyoto rats (WKY), to determine its comparative effects on the systolic blood pressure, plasma and lung angiotensin-converting enzyme (ACE) activities, as well as on lipid peroxidation in the kidneys, as measured by thiobarbituric acid (TBA) assay. This study showed that the aqueous extract of *A. paniculata* lowers systolic blood

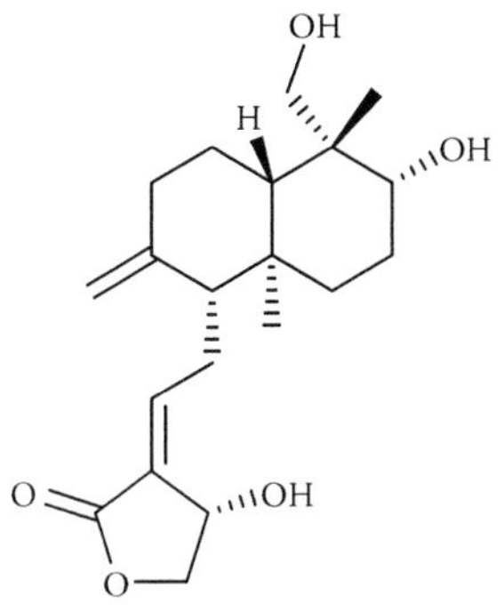

68

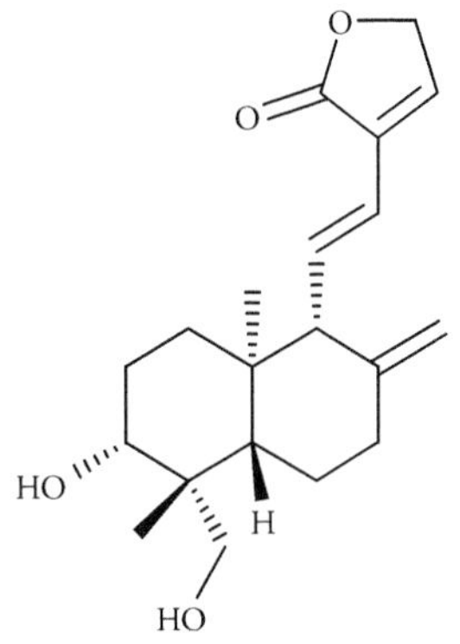

69

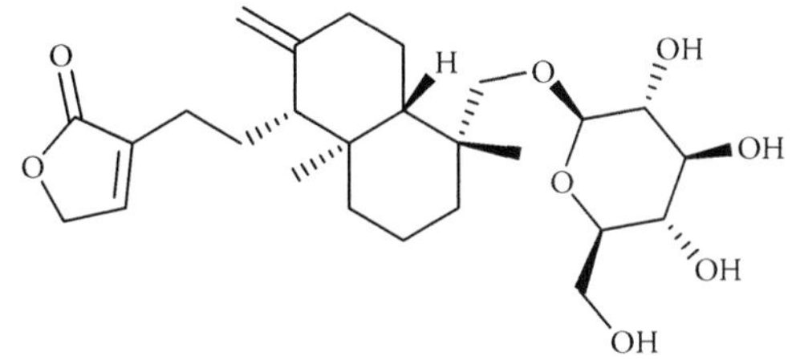

70

**FIGURE 8.2** Chemical structures of andrographolide (**68**), 14-deoxy-11,12-didehydro-andrographolide (**69**) and neoandrographolid (**70**).

pressure in the SHR, possibly by reducing circulating ACE in the plasma as well as by reducing free radical levels in the kidneys. The mechanism(s) of hypotensive action, however, seems to be different in WKY rats (Zhang and Tan, 1996).

### Chemistry

Andrographolide (**68**) is the major constituent extracted from the leaves of the plant and is a bicyclic diterpenoid lactone (Yoopan *et al.*, 2007; Chao and Lin, 2010) (Figure 8.2). The other major active diterpenoids are 14-deoxy-11,12-didehydroandrographolide (**69**) and neoandrographolide (**70**) (Yoopan *et al.*, 2007; Chao and Lin, 2010) (Figure 8.2).

## *Averrhoa belimbi* L.

*Averrhoa belimbi* L. (Figure 8.3) probably originated from eastern Indonesia but has since been introduced to the tropical regions of the world, including in Borneo, and usually cultivated for its fruit.

### Traditional Uses

The fruits and leaves of *A. belimbi* have traditionally been used in Borneo for the treatment of hypertension.

**FIGURE 8.3** Averrhoa *belimbi* tree (Photo credit: Shuttlestock).

## Biological Activity

It was demonstrated in a trial with 60 hypertensive people that tincture of *A. belimbi* fruit can lower both the diastolic and systolic blood pressure in them (Lestari *et al.*, 2018). It was also shown using left and right atria isolated from guinea pig that aqueous extract of *A. belimbi* fruit decreased both atrial contractility (inotropic) and heart rate (chronotropic) by activation of M2 receptors increasing potassium efflux, associated with decreased calcium entry in cardiomyocyte (dos Santos *et al.*, 2018).

It should, however, be noted with caution that *A. belimbi* fruit juice can cause deposition of calcium oxalate crystals in renal tubules resulting in acute oxolatteneuropathy (Nair *et al.*, 2014) as well as neurotoxicity that is attributable to carambolin (Caetano *et al.*, 2017).

## Chemistry

There have not been any bioactive compounds isolated from *A. belimbi* so far that are linked to the anti-hypertensive property.

## *Annona muricata* L.

*Annona muricata L.* is believed to have originated from tropical South America and has been introduced for its fruit throughout the tropics.

### Traditional Uses

The fruit and leaves of *A. muricata* have been used for anti-hypertensive effect.

### Biological Activity

The hypotensive effects of *A. muricata* are not mediated through muscarinic, histaminergic, adrenergic and nitric oxide pathways, but rather through peripheral mechanisms involving antagonism of $Ca^{(2+)}$ (Nwokocha *et al.*, 2012).

### Chemistry

So far, no anti-hypertensive bioactive compounds have been isolated from *A. muricata.*

## *Aquilaria microcarpa* Baill.

*Aquilaria microcarpa* Baill. is a tropical tree found in Borneo which produces a resin due to a fungal infection on the wood. This resin is highly sought after by some cultures globally for making incense and perfume (Figure 8.4).

### Traditional Uses

Gaharu tea (Figure 8.5) made from leaves of different species of *Aquilaria* has traditionally been used to lower hypertension as well as blood sugar level in Sarawak. The tea (leaves are fermented and dried) can be obtained in some Chinese herbal stores in Sarawak and is also exported to China.

**FIGURE 8.4** *Aquilaria microcarpa.* (Photo credit: S. Teo).

**FIGURE 8.5** Gaharu tea from *Aquilaria microcarpa* from a Chinese drugstore in Kuching (Photo credit: S. Teo).

## Biological Activity

The aqueous leaf extract of various species of *Aquilaria* has been shown to display anti-hypertensive activity (Wisutthathum *et al*., 2019).

## Chemistry

Gaharu tea exhibited anti-hypertensive activity through a vasorelaxant; at the same time, the tea is free of acute cytotoxicity towards vascular smooth muscle (Wisutthathum *et al*., 2019).

Gaharu tea and its main bioactive compound, mangiferin (**71**) (Figure 8.6), dilated mesenteric arteries and aortae *in vitro*. AE dilated rat mesenteric arteries ($EC_{50}$ ~ 107 µg/ml, $E_{max}$ ~ 95%) more than aorta ($EC_{50}$ ~ 265 µg/ml, $E_{max}$ ~ 76%,

**FIGURE 8.6** Structure of mangiferin (**71**).

$p < 0.05$) (Wisutthathum *et al.*, 2019). Wisutthathum *et al.* (2019) noted the opposite with mangiferin (1–100 μM), which dilated the mesenteric artery more potently than the aorta while the maximum relaxation was lower than that observed with aqueous extract (41% in the mesenteric artery and <10% in the aorta). Isolated vascular smooth muscle cells incubated in mangiferin for an hour also showed no cytotoxicity (Wisutthathum *et al.*, 2019).

## *Cordyline fruticosa* (L.) A.Chev.

*Cordyline fruticosa* (L.) A.Chev. (Figure 8.7) is a small shrub that belongs to the Asparagaceae family and has cultural significance amongst some ethnic groups in Borneo.

### Traditional Uses

The Lun Bawang of Sarawak boil the plant together with a wild species of ginger to be drunk regularly to lower hypertension (Chai, 2000).

### Biological Activity

There are no known studies to screen *C. fruticosa* for anti-hypertensive activity so far.

**FIGURE 8.7** *Cordyline fruticosa* (Photo credit: S. Teo).

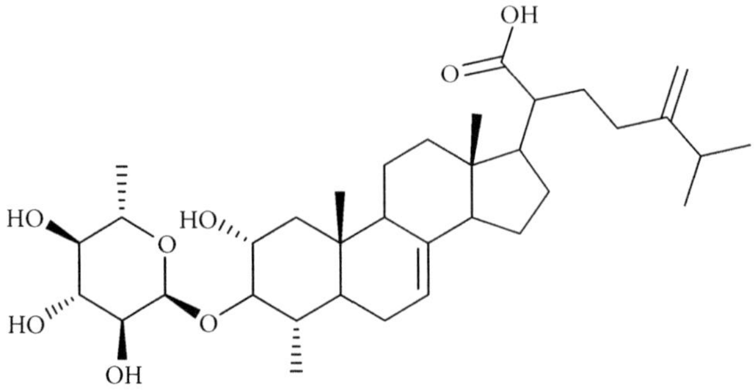

**FIGURE 8.8** Fruticoside C (**72**), one of the steroidal saponins isolated from *Cordyline fruticosa.*

## Chemistry

There is also no known bioactive anti-hypertensive compound isolated from *C. fruticosa* that has been directly linked with anti-hypertensive activity so far. Steroidal saponins such as fructicoside C (**72**) (Figure 8.8) have been isolated from *C. fruticosa* (Fouedjou *et al.*, 2014; Ponou *et al.*, 2019) and saponins have been demonstrated to exhibit anti-hypertensive activity (Manivannan *et al.*, 2015; Singh and Chaudhur, 2018).

# *Bauhinia purperea* L.

*Bauhinia purperea* L. (Figure 8.9) belongs to the Leguminosae family and is sometimes planted as an ornamental plant.

## Traditional Uses

A tea made from the roots of *Bauhinia purperea* is consumed for hypertension.

## Biological Activity

There have been no studies done to demonstrate that *B. purperea* has anti-hypertensive activity. However, the aqueous extract of another *Bauhinia* species, *B. forficata*, exhibited anti-hypertensive activity in rats. This activity has been attributed to the releasing of nitric oxide (dos Anjos *et al.*, 2013).

## Chemistry

No studies have been done to attribute the anti-hypertensive effect in *Bauhinia* to a bioactive compound including the triterpene saponin obtained from *B. variegata* (Mohamed *et al.*, 2009).

# *Colocasia esculenta* L.

*Colocasia esculenta* L. (Figure 8.10) or coco yam is a cultivated crop, and the rhizome is edible and a source of starch.

**FIGURE 8.9** *Bauhinia purperea* (Photo credit: S. Teo).

## Traditional Use

The indigenous community in Sintang, West Kalimantan boil the whole plant and drink the aqueous extract to treat hypertension.

## Biological Activity

In a study to evaluate the hypertensive effect of an ethanolic extract of *C. esculenta* (EECE) using male Sprague Dawley rats, it was found that 40 mg of EECE/200 g body weight per day significantly ($p < 0.05$) decreased 16.07% and 13.67% of the systolic and diastolic blood pressures, respectively (Prastiwi *et al.*, 2016b). In a study using renal artery-occluded hypertensive rats and noradrenalin-induced hypertension in rats, it was found to possess anti-hypertensive activity which may be attributed to ACE inhibitory, vasodilatory, $\beta$-blocking, and/or $Ca^{2+}$ channel blocking activities (Vasant *et al.*, 2012).

## Chemistry

There has been no work done to isolate the anti-hypertensive compound(s) from *C. esculenta* thus far.

**FIGURE 8.10** *Colocasia esculenta* (Photo credit: S. Teo).

# 9 Anti-Diarrhoeal Plants

*Irawan Wijaya Kusuma*

## CONTENTS



## INTRODUCTION

The World Health Organization (WHO) defined diarrhoea as the passage of 3 or more loose or liquid stools per day, or more frequently than is normal for the individual. As a leading killer of children, UNICEF reported that diarrhoea caused approximately 8% of all deaths among children under age 5 worldwide in 2016. Nearly 1.7 billion cases of childhood diarrhoeal diseases are reported every year.

Rotavirus and Norwalk-like viral infections cause most diarrhoeal diseases, particularly in young children, whilst microorganisms such as *Salmonella, Shigella, Campylobacter, Giardia, Cryptosporidium, Escherichia coli, Vibrio cholera, Staphylococcus aureus* and *Clostridium difficile* are the most common pathogenic bacteria isolated from the diarrhoeal patients. Gastrointestinal infection, particularly due to enterotoxigenic *E. coli* (ETEC), was reported to be responsible for health problems among children in developing countries.

Based on the biological activity and their active compounds, many plants have been identified as good sources for natural anti-diarrhoeal drugs.

**FIGURE 9.1** Leaves (left) and bulbs Bulbs of *Eleutherine americana* (Photo credit: I.W. Kusuma).

## *Eleutherine americana* (L.) Merr.

*Eleutherine americana* L. (Aubl.) Merr. (Figure 9.1), known locally as bawang tiwai, is a plant introduced from tropical regions of the Americas and widely distributed in Borneo, especially in West Kalimantan, and is widely used by the Dayak, an indigenous group in Borneo. In some regions, the plant is also known as Dayak onion.

### Traditional Uses

Indigenous tribes in Borneo use this plant as a medicine for various diseases. Bulbs that are elliptical and red are used as diuretics, laxatives and vomiting treatments. The bulbs are soothing and useful for inflammation of the intestines, constipation and dysentery.

The leaves are used as medicine for bloody diarrhoea and the pulp is applied externally as a paste. The leaves are also useful for treating fever and nausea. The leaves are crushed together with a sprinkling of other herbs taken by the puerperal woman.

### Biological Activity

*n*-Hexane, ethyl acetate and ethanolic extracts of the *E. americana* bulb were reported to inhibit Gram-positive and Gram-negative bacteria such as *Bacillus cereus, Shigella sp.* and *Pseudomonas aeruginosa*. Extensive research on antibacterial activity on this plant showed that the ethanolic extract of the bulb of the plant had potent activity against *Campylobacter* spp. (Sirirak and Voravuthikunchai, 2011).

### Chemistry

Studies on the phytochemicals of *E. americana* reported that a large number of bioactive compounds can be classified into three main groups, i.e. naphthalene, anthraquinone and napthoquinone (Figure 9.2). Some of the bioactive compounds

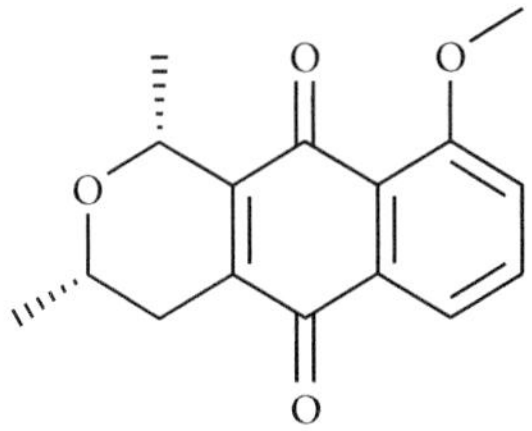

73

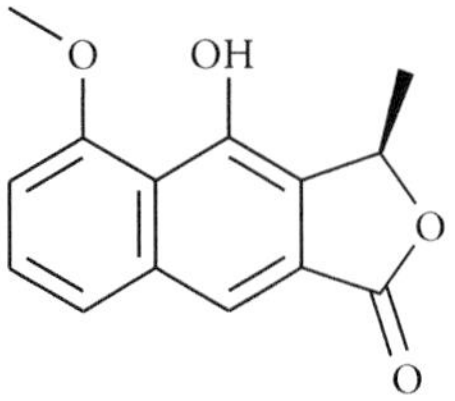

74

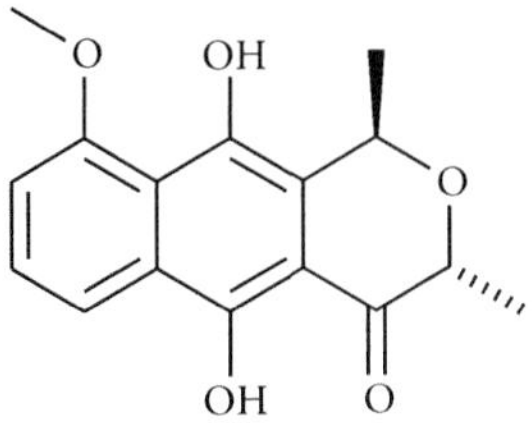

75

**FIGURE 9.2** Three naphthoquinone compounds from *Eleutherine americana* – hongconin (**73**), eleutherol (**74**) and eleutherin (**75**).

from *E. americana* were hongconin (**73**), eleutherol (**74**), isoeleutherol, eleutherin (**75**), isoeleutherin, eleutherinol, eleutherinoside A&B, elecanacin, eleuthinone A, eleuthraquinone A&B, eleucainorol, erythrolaccin, (–)-3-[2-(acetyloxypropyl)]-2-hydroxy-8-methoxy-1,4-naphtoquinone, 1,5 dihydroxy-3-methyl anthraquinone, 3,4,8-trihydroxy-1-methyl-anthra-9,10-quinone-2-carboxylic acid methyl-ester, 3,4-dihydro-1,3-dimethyl-1-H-naphthol (2,3) pyran-5,10,dione, 4,8-dihydroxy-3,4-dimethoxy-1-methyl-anthraquinone-2-carboxylic acid methyl ester, 3,2-acetyloxy prophyl-2-hydroxy-8-methoxy-1,4-napthoquinone, 1,2-hydroxy-8-methoxy-3methy lantraquinone, 9-hydroxy-8-methoxy-1-methyl-1, 3-dihydronaphtho [2, 3-c] furan-4-*O*-β-D-glucopyranoside, 1,2-dihydroxy-8-methoxy-3-methyl-anthra-9,10-quinone and 9-methoxy-1, 3-dimethyl-1H-naphtho [2, 3-c] pyran-5, 10-dione (Zhengxiong *et al.*, 1986; Alves *et al.*, 2003; Paramapojna *et al.*, 2008; Isanu *et al.*, 2014; Chena *et al.*, 2019).

Hongconin (**73**), eleutherol (**74**), eleutherin (**75**), isoeleutherin, anthraquinones, and elecanacin (Figure 9.2) present in the ethanolic extracts of *E. americana* were reported to inhibit enzymes that play an important role in the production of enterotoxin by *S. aureus* in food (Ifesan *et al.*, 2010; Ifesan *et al.*, 2009).

## *Melastoma malabathricum* L.

*Melastoma malabathricum* L. (Figure 9.3), known locally as karamunting or senduduk, is a plant species usually found in secondary forests. The plant is found

**FIGURE 9.3** Leaves (above) and fruits (below) of *Melastoma malabathricum* (Photo credit: I.W. Kusuma).

throughout Borneo, including Kalimantan where its existence causes a serious problem due to its invasive nature.

## Traditional Uses

*M. malabathricum* is believed to be able to neutralize poison. The parts used are leaves, fruits, seeds and roots. In addition, the plant can also be used to treat several types of diseases such as digestive disorders (dyspepsia), bacillary dysentery, diarrhoea, hepatitis, leucorrhoea, defecated blood and inflammation of blood vessel walls.

## Biological Activity

Flavonoids from the leaves of *M. malabathricum* were reported to possess antibacterial activity against diarrhoeal-causing strains such as *E. coli* and *B. cereus* (Omar *et al.*, 2013).

## Chemistry

Phytochemical composition and pharmacological effects of *M. malabathricum* leaves were well investigated. Preliminary screening of *M. malabathricum* leaves indicated the presence of flavonoids, steroids, triterpenoids, tannins and quinine. Furthermore, 95% ethanol extract contained antibacterials from the flavonoid group, such as quercetin (**44**), quercitrin (**76**) (Figure 9.4) and myricetin in the form of glycosides, as well as phenolic acid derivatives such as *p*-hydroxybenzoic acid and *p*-coumar acid in ester form and terpenoids such as α-amyrin (**77**) (Wong *et al.*, 2012).

GC-MS analysis of the leaves of *M. malabathricum* revealed the occurrence of 5-hydroxymethylfurfural, pyrogallol, phytol, palmitic acid methyl ester, 8,11-

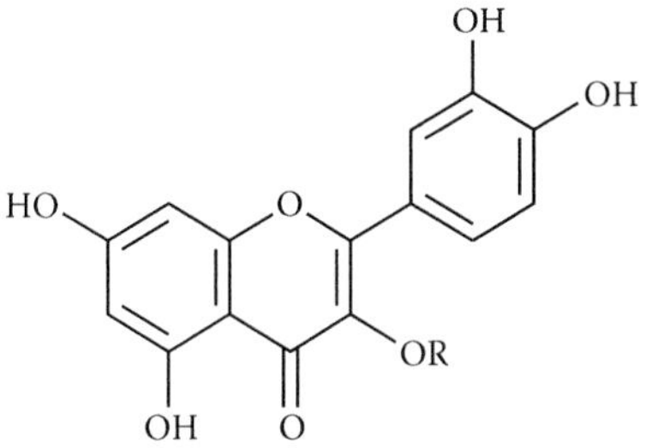

76

74 R = H

75 R = Rhamnose

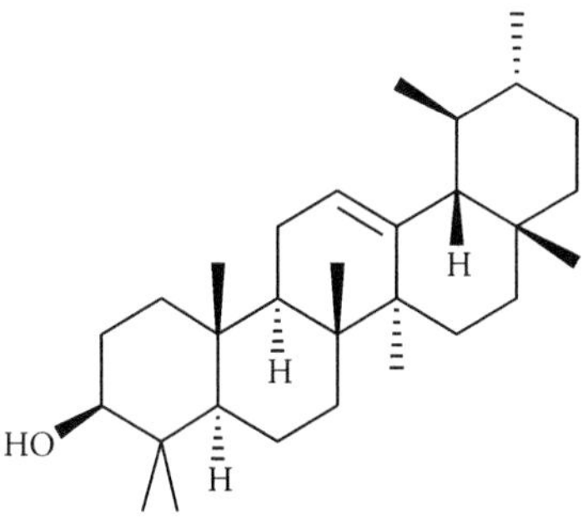

**FIGURE 9.4** Three active compounds from *Melastoma malabathricum* – quercetin (**44**), quercitrin (**76**) and α-amyrin (**77**).

octadecadenoic acid, stearic acid methyl ester, trans-squalene and tocopherol. Isolation of phytochemicals from chloroform and ethyl acetate fractions of *M. malabathricum* showed the presence of ursolic acid, 2α-hydroxyursolic acid, asiatic acid, β-sitosterol 3-*O*-β-D-glucopyranoside, glycolipid glycerol 1,2-dilinolenyl-3-*O*-β-D-galactopyanoside, kaempferol, kaempferol 3-*O*-α-L-rhamnopyranoside, kaempferol 3-*O*-β-D-glucopyranoside, kaempferol 3-*O*-β-D-galactopyranoside, *kaempferol* 3-*O*-(2″,6″-di-*O*-E-*p*-coumaryl)-β-D-galactopyranoside, quercetin and ellagic acid (Joffry *et al.*, 2012).

## *Rhodomyrtus tomentosa* (Aiton) Hassk.

*Rhodomyrtus tomentosa* (Aiton) Hassk. (Figure 9.5) is a species in the genus of *Rhodomyrtus* in Myrtaceae family. *R. tomentosa* is native to South and Southeast Asia and widely distributed throughout Borneo. The spread of this plant has been a serious problem as it is an invasive plant species.

### Traditional Uses

*R. tomentosa* has been used in diarrheoal and wound treatments in Borneo. Fruits of *R. tomentosa* are eaten raw to treat diarrheoa.

### Biological Activity

Ethanolic extract of the plant showed good activities against some Gram-positive bacteria. Ethanolic extract and an acylphloroglucinol compound isolated from the *R. tomentosa* leaves have been reported to have profound antibacterial activities. *n*-Hexane, ethyl acetate and methanol extracts of the leaves of *R. tomentosa* displayed moderate activity against several *S. aureus* and *E. coli* with $EC_{50}$ between 144 and 262 ppm whilst fruit and stem extracts showed $EC_{50}$ between 64–782 ppm and $EC_{50}$ between 43 and 189 ppm, respectively (Bach *et al.*, 2018; Lavanya *et al.*, 2012).

**FIGURE 9.5** Leaves and flowers (above) and fruits (below) of *Rhodomyrtus tomentosa* (Photo credit: I.W. Kusuma).

Some phloroglucinols isolated from *R. tomentosa* exhibited interesting antibacterial activities. Rhodomyrtone (**78**) isolated from the leaves of the plant showed potent activity against *E. coli* and *S. aureus* (Saising *et al.*, 2011; Srisuwan *et al.*, 2018) and has generated a great interest in the development of new antibiotics derived from natural sources.

Rhodomyrtone [6,8-dihydroxy-2,2,4,4-tetramethyl-7-(3-methyl-1-oxobutyl)-9-(2-methylpropyl)-4,9-dihydro-1H-xanthene-1,3(2H)-di-one] (**78**) from *R. tomentosa* have been reported to have potent activity against Gram-positive bacteria including *B. cereus*, *Bacillus subtilis*, *Enterococcus faecalis*, *S. aureus* and methicillin-resistant *S. aureus* (MRSA). Rhodomyrtone (**78**) was also found to have the ability to inhibit bacteria activity by preventing staphylococcal biofilm formation as well as disrupting the mature biofilms (Mordmuang *et al.*, 2015).

Due to the excellent antibacterial nature of rhodomyrtone (**78**), the antibacterial mechanisms of this compound have been well investigated. The unique and novel

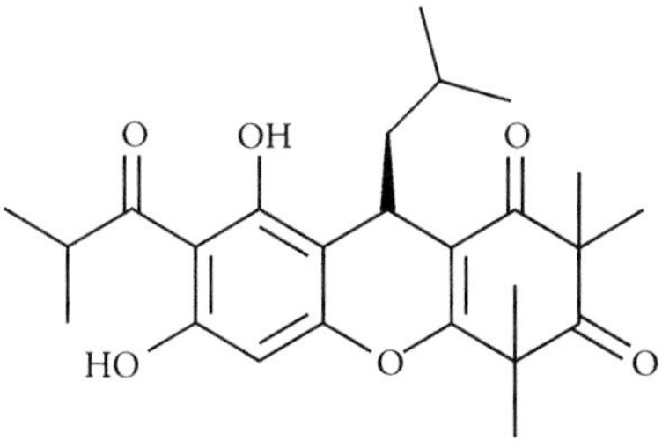

**FIGURE 9.6** Rhodomyrtone (**78**), a promising antibacterial candidate from *Rhodomyrtus tomentosa.*

antibacterial mechanisms, as opposed to those of existing cell membrane-targeting drugs, minimizing of cross-resistance risk as well as significant activity against antibiotic resistance *S. aureus* made rhodomyrtone (**78**) an extremely interesting antibacterial drug candidate (Saising *et al.*, 2011).

### Chemistry

Chemical analysis on *R. tomentosa* leaves showed the presence of lupeol, β-amyrin, β-amyrenonol, botulin, tomentosin, peduculagin, casuarin, castalagin, myricetin derivatives and rhodomyrtone (**78**) (Figure 9.6). The stem of the plant contains friedelin, lupeol, taraxerol, betulin and betulin-3-acetate whilst the flower contains malvidin-3-glucoside, pelargonidin-3,5-biglucoside, cyaniding-3-galactosie and delphinidin-3-galactoside and the bark and twigs contain combretol (Saising *et al.*, 2011).

## *Alstonia scholaris* (L.) R.Br.

*Alstonia scholaris* (L.) R.Br. (Figure 9.7), known locally as pulai, is a species in the family Apocynaceae. This plant is indigenous to Borneo and is fast growing (Adinugraha, 2011). *A. scholaris* has some uses such as for making crates, matches, crafts and pulp. Pulai is widely distributed in Borneo and commonly found in the secondary forests from the lowlands to an elevation of about 1000 m above sea level.

### Traditional Uses

*A. scholaris* is a tropical plant having rich medicinal properties. Indigenous tribes in Kalimantan use the leaves to treat various diseases such as diarrhoea, dysentery, malaria and snake bites. The Bahau tribes in North Kalimantan use the stem bark of the plant to treat diarrhoea.

The bark is the most intensively used part of the plant and is used in many herbal formulas. It is a bitter tonic and febrifuge and is reported to be useful in the treatment of malaria, diarrhoea and dysentery.

### Biological Activity

The stem bark extracts of *A. scholaris* have been tested against several diarrhoeal-causing bacterial strains. Among methanolic, aqueous and alkaloid extracts, the aqueous extract displayed the best activity against Gram-positive and Gram-negative bacteria including *B. subtilis* and *E. coli* (Khyade and Vaikos, 2009b).

**FIGURE 9.7** Leaves and flowers of *Alstonia scholaris* (Photo credit: I.W. Kusuma).

## Chemistry

Intensive phytochemical investigation revealed that *A. scholaris* possessed nearly 400 chemical substances including alkaloids, iridoids, coumarins, flavonoids, leucoanthocyanins, phenolics steroids, saponins and tannins. Some pentacyclic-triterpenoids were reported to have antibacterial activity against *B. cereus* and *S. aureus*, two diarrhoeal-causing bacterial strains (Limsuwan *et al.*, 2009).

One of the most important chemical types found in *A. scholaris* is the alkaloids, and this plant was reported to contain at least 180 alkaloids. However, the investigation into their biological activities is rather limited (Molly *et al.*, 2011). Alkaloids such as alstonidine, *O*-methylmacralstonine, macralstonine *O*-acetylmacralstonine, alstonine, ditamine, echicaoutchin, corialstonidine, corialstoninechlorogenine, villalstonine, pleiocarpamine, villalstonine, macrocarpamine and others have been isolated from the bark of the plant (Khyade and Vaikos, 2009a).

Recent study revealed that some new alkaloids possessed significant antibacterial activities in comparison to that of commercial antibiotics. Scholarisinine-type compounds are a new type of alkaloids (Figure 9.8) such as scholarisinine T (**79**),

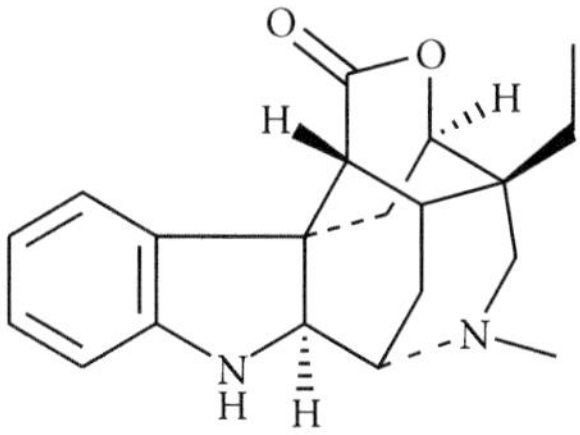

79

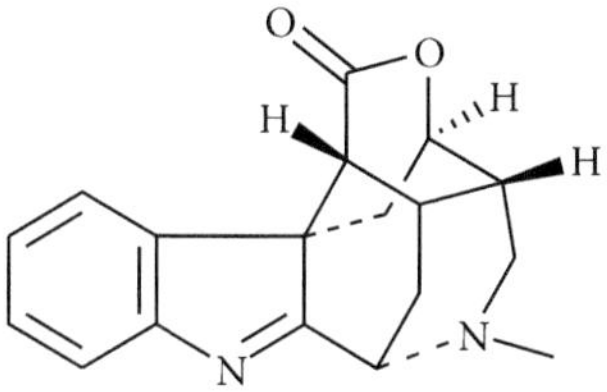

80

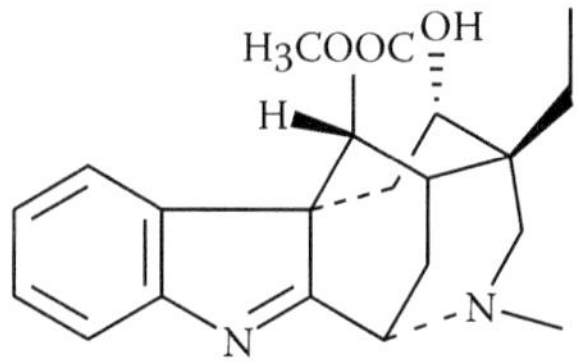

81

**FIGURE 9.8** Three antibacterial compounds from A. scholaris – Scholarisinine T (**79**); Scholarisinine U (**80**); Scholarisinine (**81**) active against *S. aureus* and *E. coli.*

scholarisinine U (**80**) and scholarisinine (**81**) isolated from *A. scholaris* and were found to be highly effective in inhibiting the growth of diarrhoeal-causing bacteria (Yu *et al.*, 2018).

## Other Anti-Diarrhoeal Plants

In a comprehensive ethnopharmacological study of tropical medicinal plants traditionally used by indigenous tribes in Borneo during 2012–2015, hundreds of medicinal plants were collected and compiled for their folkloric uses. Some plants were found to be used by local people to treat stomach diseases, including diarrhoea, dysentery and bloody faeces.

*Tetracera indica*, *Macaranga hypoleuca* and *Callicarpa longifolia* were reported by Bahau Dayak in East Kalimantan to be effective for the treatments of diarrhoea and dysentery. *Syzygium aromaticum*, an aromatic tree species, was reported by the Bulungan tribe in eastern part of Borneo to be commonly used to cure stomach diseases including diarrhoea. A well-known plant already developed as an herbal drug for diarrhoea treatment in Indonesia is *Psidium guajava*. The herbal products are available in the form of dry leaves, encapsulated leaf powder and tablets. Recent investigation has revealed that *P. guajava* leaf extract was effective in clearing diarrhoeal infection in mice, the model suggesting the potential of the plant for use in human diarrhoeal treatment (Gupta and Birdi, 2015).

# 10 Plants for Eye-Related Diseases

*Stephen P. Teo*

## CONTENTS

## INTRODUCTION

Infection of the eyes, also known as conjunctivitis, is a frequent and potentially serious problem. The infections can be due either viral, bacterial or fungal. Viral conjunctivitis is mainly caused by adenovirus (65–90%) and to a lesser extent by herpes simplex virus. On the other hand, conjuctivitis caused by bacteria are mainly due to infections by *Staphylococcus aureus*, *Haemophilus influenzae*, *Streptococcus pneumonia* and *Pseudomonas aeruginosa*. Fungal infections are quite rare. Most of the medicinal plants used to treat conjunctivitis are administered as eye drops. Some of these plants might have antiviral, antibacterial and antifungal activities as well as anti-inflammatory activity and can also be used as an eye wash to clear the eye or after injury.

### Polyalthia cauliflora Hook. f. & Thomson var. beccarii (King) J. Sincl. and Polyalthia flagellaris Becc. (Airy Shaw)

*Polyathia cauliflora* Hook. f. & Thomson var. *beccarii* (King) J. Sincl. and *Polyathia flagellaris* Becc. (Airy Shaw) are two species that belong to the family Annonaceae). They are often under-storey trees in the tropical rainforests of Borneo.

#### Traditional Uses

The sap from warmed fresh leaves of both *P. cauliflora* var. *beccarii* and *P. flagellaris* (Figure 10.1) are used to treat sore eyes by the Ibans of Sarawak.

#### Biological Activity

There have been no studies done on either *P. cauliflora* var. *beccarii* or *P. flagellaris* so far. However, there were studies done on other *Polyalthia* species elsewhere.

Species of *Polyalthia* have been known to possess antibacterial and antiviral activities. It has been shown that the methanolic leaf extract of *Polyalthia longifolia* offers promising antiviral activity against paramyxoviruses and acts by inhibiting the entry and budding of viruses (Yadav *et al.*, 2019). However, no work has been done on adenovirus.

A *Polyalthia* species has been shown to have a protective effect against the formation of cataracts on the eyes. However, ethanol and chloroform extracts of *P. longifolia* have been shown to possess glucose-induced cataractogenesis activity in goat lenses (Sivashanmugam and Chatterjee, 2012). *Polyalthia* species are also known to have anti-inflammatory activity (Yao *et al.*, 2019).

#### Chemistry

Two antiviral acetylenic fatty acid compounds 19-(2-furyl)nonadeca-5,7-diynoic acid (**82**) and 19-(2-furyl)nonadeca-5-ynoic acid (**83**) (Figure 10.2) – have been isolated from the roots of *P. evecta* and both have demonstrated activity against herpes simplex type 1 virus (Kanokmedhakul *et al.*, 2006).

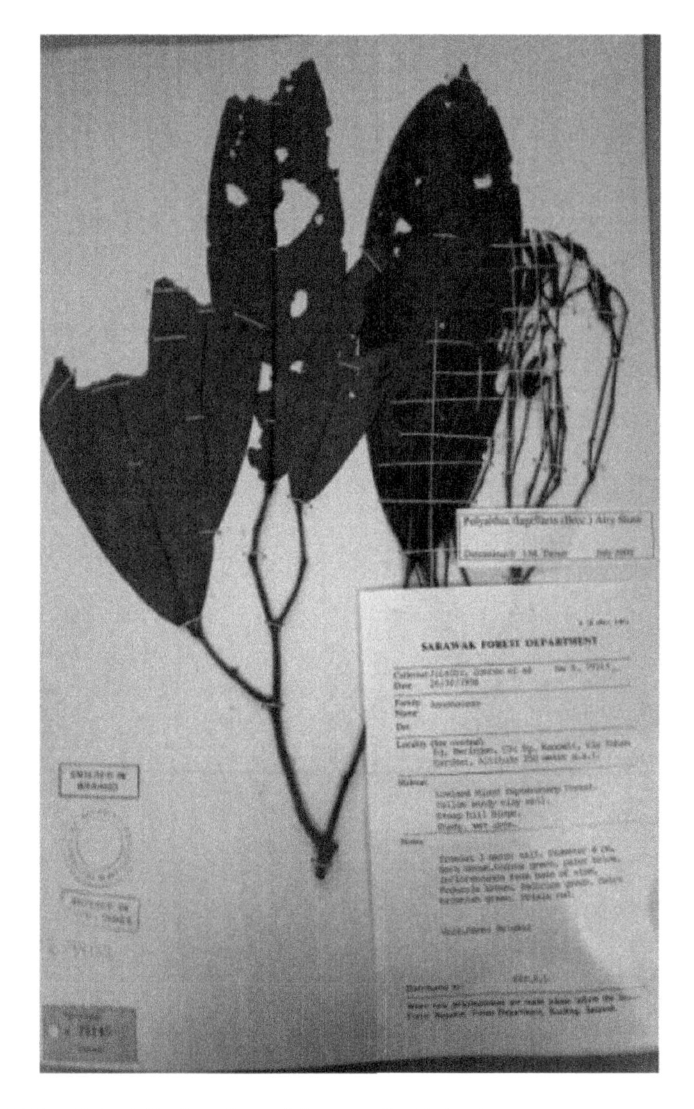


**FIGURE 10.1** *Polyalthia flagellaris* (Photo credit: S. Teo).

## *Mitrephora glabra* Scheff.

*Mitrephora glabra* Scheff. (Figure 10.3) is a tree belonging to the Annonaceae family and is endemic to Borneo.

### Traditional Uses

The Iban of Sarawak made use of the pounded fresh leaves as a remedy for sore eyes by applying the sap as an eye drop.

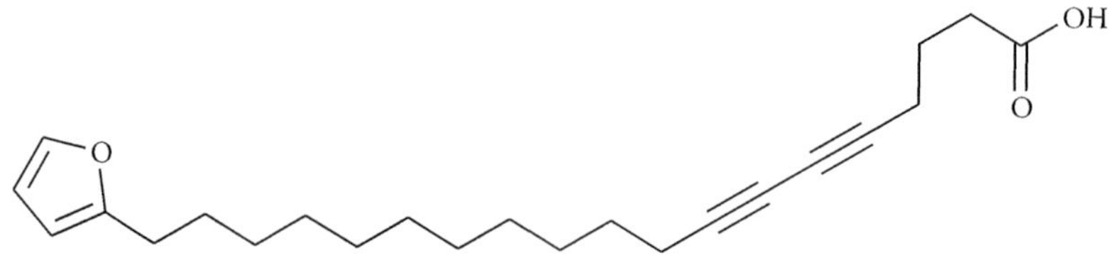

**83**

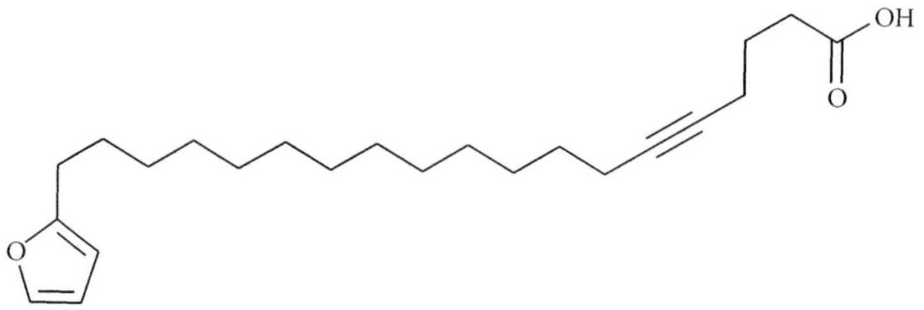

**84**

**FIGURE 10.2** Antiviral acetylenic fatty acid compounds – 19-(2-furyl)nonadeca-5,7-diynoic acid (**83**) and 19-(2-furyl)nonadeca-5-ynoic acid (**84**).

**FIGURE 10.3** Fruits of *Mitrephora glabra* (Photo credit: S. Teo).

## Chemistry and Biological Activity

Zgoda *et al.* (2001) isolated two polyacetylene carboxylic acids, namely 13,14-dihydrooropheic acid (**84**) and oropheic acid (**85**) (Figure 10.4), from the stem bark of *M. glabra* and both compounds displayed significant activities against both methicillin-resistant *S. aureus* and *Mycobacterium smegmatis*.

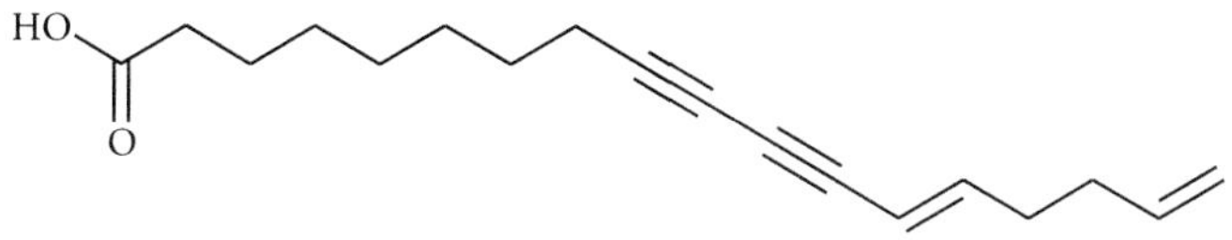

HO O 85

HO O 86

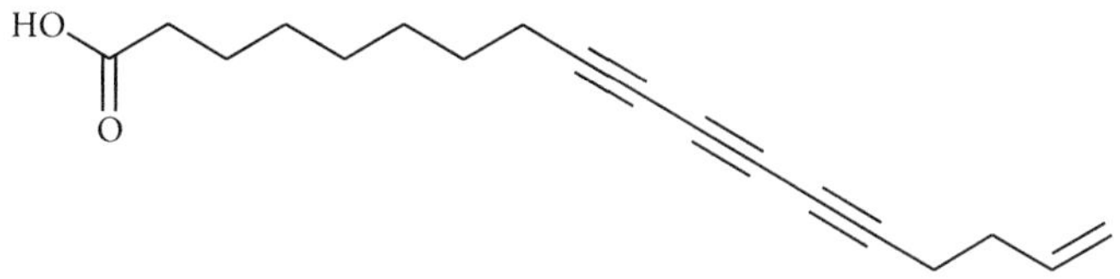

**FIGURE 10.4** Polyacetylene carboxylic acids – 13,14-dihydrooropheic acid (**85**) and oropheic acid (**86**).

Three *ent*-kaurenoids, five polyacetylenic acids/esters and one aporphine alkaloid, liriodenine, were isolated from the stem bark of *M. glabra* (Li *et al.*, 2009) which showed activities against a Gram-positive bacteria (*Micrococcus luteus*), a mycobacterium (*Mycobacterium smegmatis*), a yeast (*Saccharomyces cerevisae*) and a fungus (*Aspergillus niger*) with the best activities shown by polyacetylenic acids/esters and the alkaloid. Li *et al.* (2005) isolated three *ent*-trachylobanediterpenoids and all showed equipotent activities against a Gram-positive bacteria (*M. luteus*), a mycobacterium (*M. smegmatis*) and a yeast (*S. cerevisiae*).

## *Imperata cylindrica* (L.) Raeusch

*Imperata cylindrical* (L.) Raeusch or Cogon grass (Figure 10.5) is a noxious pantropical and subtropical weed from Asia, Australia and the Pacific islands, and later introduced to the Americas.

### Traditional Uses

In West Kalimantan, the shoot of *I. cylindrical* was squeezed and the sap was used as eye drops for sore eyes (Meliki *et al.*, 2013).

### Biological Activity

*I. cylindrica* was shown to be rich in phenolic compounds, especially flavonoids (Qaddoori *et al.*, 2016). Flavonoids have been demonstrated to display antiviral, antibacterial and antifungal activities (Moreno *et al.*, 2006). Moreno *et al.* (2006) reported that the antibacterial action of quercetin and other phenolic compounds might be related to inactivation of certain cellular enzymes proteins, which are responsible for bacterial membrane formation, by forming irreversible complexes. Xie *et al.*

**FIGURE 10.5** *Imperata cylindrica* (Photo credit: S. Teo).

(2014) proposed numerous antibacterial mechanisms of flavonoids which involve the inhibition of nucleic acid synthesis, cytoplasmic membrane function, energy metabolism, the attachment and biofilm formation, porin on the cell membrane as well as a change in membrane permeability, and attenuation of the pathogenicity.

### Chemistry

Quercetin (**44**) (Figure 10.6) has been shown to display excellent antibacterial activities against Gram-positive and Gram-negative bacteria as well as the drug-resistant bacteria, MRSA (Ismail *et al.*, 2011; Parkavi *et al.*, 2012; Qaddoori, 2016). Apart from antibacterial activities, quercetin (**44**) also demonstrated antiviral (Johari *et al.*, 2012) and antifungal (Rocha *et al.*, 2019) activities (Wang *et al.*, 2012) as well as anti-inflammatory activity (Choi *et al.*, 2012).

## *Dicranopteris linearis* (Burm. f.) Underw.

*Dicranopteris linearis* (Burm. f.) Underw. (Figure 10.7) belongs to the family of Gleicheniaceae and is a common creeping fern. It is found in disturbed areas in the rainforests or by hillsides, often forming thickets.

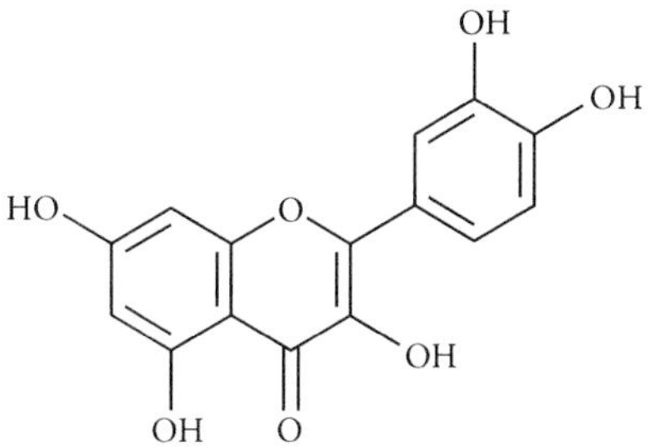

**FIGURE 10.6** Quercetin (**76**).

**FIGURE 10.7** Thickets of *Dicranopteris linearis* (Photo credit: S. Teo).

## Traditional Uses

Sap squeezed from the stalk of the fern fiddlehead is used by the Kenyah of East Kalimantan as eye drops to sooth pain after eye injury (Leaman, 1996).

## Biological Activity

*D. linearis* showed mild *in vitro* antibacterial activity. In a study using petroleum ether, acetone, methanolic and aqueous extracts of *D. linearis*, only acetone and methanolic extracts showed antibacterial activity with acetone displaying activity with an MIC of 8 and 16 mg/mL against *M. luteus* and *S. aureus,* respectively (Toji *et al.*, 2007). In another study using aqueous, chloroform and methanolic extracts of the leaves of *D. linearis*, only the methanolic extract displayed antibacterial activity (Zakaria *et al.*, 2010). Further fractionation of the methanolic extract gave 3 fractions with MIC/MBC values ranging between 1250 and 2500 µg/mL against the methicillin-resistant (*S. aureus* 33591) and the susceptible strain (*S. aureus* 25923) (Zakaria *et al.*, 2010).

For anti-biofilm activity, the aqueous and hexane fractions were the most effective for biofilm inhibition activity and biofilm disruption, respectively, when tested against biofilm of five *S. aureus* strains. The aqueous fraction demonstrated biofilm inhibition at 0.31–2.5 mg/mL while the hexane fraction displayed biofilm disruption activity at 0.07–5 mg/mL (Christina Injai). Neither fraction inhibited cell growth but rather the anti-biofilm effect was only due to the biofilm structure itself, which was confirmed by electron microscopy (Christina Injai, 2017). The hexane fraction was able to disrupt about 42–75% of *S. aureus* biofilms. The bioactive compound from *D. linearis* was identified as alpha-tocopherol which exhibited biofilm disruption activity against *S. aureus* biofilms at 0.01–0.5 mg/mL (Christina Injai, 2017).

From the time killing assay, minimum inhibitory concentration (MIC) assay and scanning electron microscopic images, a standardized fraction of methanolic extract of *D. linearis* was found to exert bacteriostatic effects against *S. aureus* and *P. aeruginosa*. Moreover, the standardized fraction was also able to potentiate the antibacterial effect of few conventional antibiotics, namely, chloramphenicol, penicillin G and ampicillin, against *S. aureus* and MRSA (Ponnusamy, 2016). Furthermore, the standardized fraction was able to enhance the proliferation of fibroblast cells and induce cell migration in the scratch-wound assay (Ponnusamy, 2016). In addition, it protected the cells against hydrogen peroxide induced oxidative stress and was non-cytotoxic towards both human and mouse fibroblast cell lines up to a concentration of 500 µg/mL (Ponnusamy, 2016).

*D. linearis* has not been screened against fungi except for plant fungal pathogens such as *A. niger*, *Macrophomina phaseolina*, *Curvularia lunata* and *Aspergillus flavus* (Devi *et al.*, 2015), whilst no studies have been done in screening against viruses. The aqueous extract of *D. linearis* was shown to exhibit anti-inflammatory activity in an experimental animal study (Zakaria *et al.*, 2008).

### Chemistry

Extracts of *D. linearis* are known to be polyphenol-rich, especially flavonoid glycosides (Figure 10.8). Amongst the flavonoid glycosides isolated from *D. linearis* were the quercitrin (**76**), afzelin (**86**), isoquercitrin (**87**) as well as a flavonol 3-*O*-glycoside (**88**) (Raja *et al.*, 1995).

## *Tetrastigma scandens* (L.) Merr.

*Tetrastigma scandens* (L.) Merr. (Figure 10.9) is a member of the climber family Vitaceae. The plant is a liana that climbs with tendrils and has palmately compound leaves.

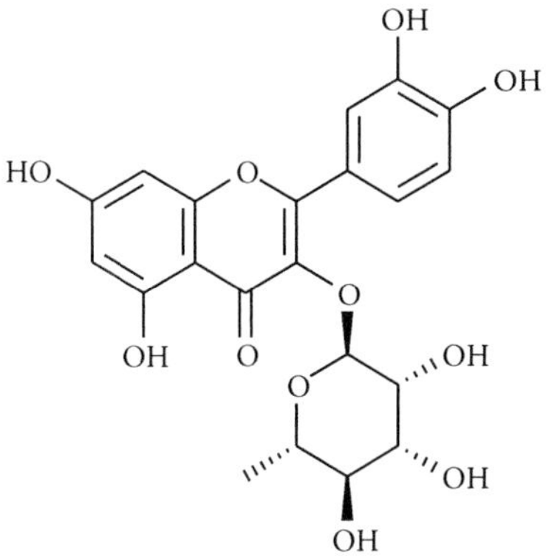

**FIGURE 10.8** Flavonoid glycosides isolated from *Dicranopteris linearis* – quercitrin (**76**), afzelin (**86**), isoquercitrin (**87**) and flavonol-3-*O*-glycoside (**88**).

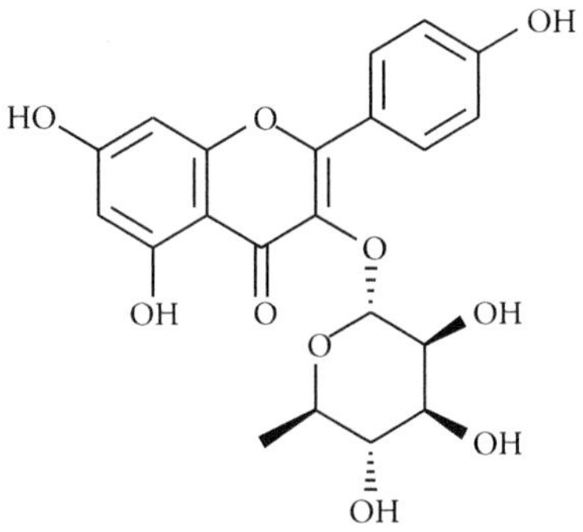

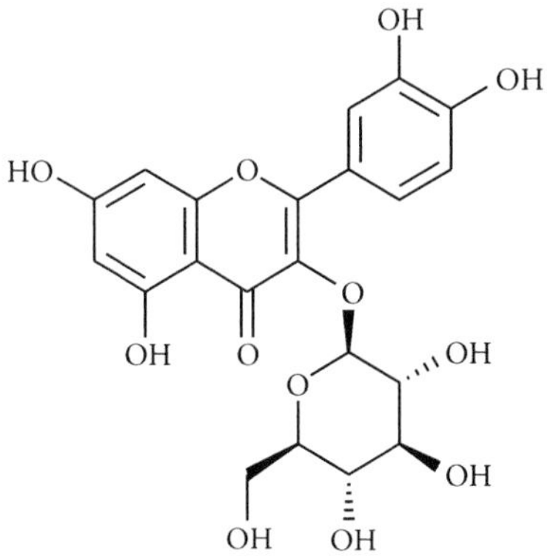

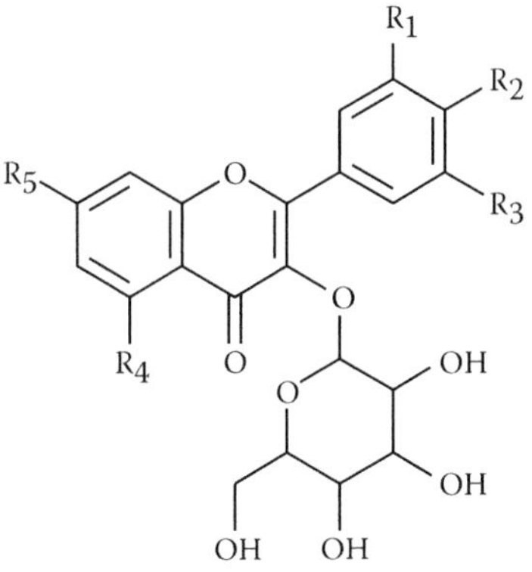

## Traditional Uses

*T. scandens* is used to sooth irritation, clean debris and clear eye vision by the Kenyah of East Kalimantan (Leaman, 1996).

## Biological Activity

*T. scandens* has not been screened against microorganisms or for anti-inflammatory activities so far.

## Chemistry

Among the compounds isolated from *T. scandens* are flavonoids and beyrane-type diterpenoids (Manh *et al.*, 2017). From the ethanolic extract of leaves of *T. scandens* L, six known flavonoid compounds have been isolated quercetin-3-*O*-

**FIGURE 10.9** *Tetrastigma scandens* (Photo credit: S. Teo).

L-rhamnoside, genistein, quercetin, quercetin 3-*O*-β-D-glucuronide, kaempferol and quercetin-3-*O*-α-L-arabinofuranoside (Thanh *et al.*, 2015). The beyerane-type diterpenoids isolated were *ent*-beyer-15-en-17-ol-19-oic acid, along with five known *ent*–kauranes, *ent*-18-acetoxy-7α-hydroxykaur-16-en-15-on, *ent*-(16S)-18-acetoxy-7α-hydroxykaur-16-en-15-on, *ent*-1α,7β-diacetoxy-14α-hydroxykaur-16-en-15-on, *ent*-1α,14α-diacetoxy-7β-hydroxykaur-16-en-15-on and *ent*-1α-acetoxy-7β, and -14α-dihydroxykaur-16-en-15-on (Manh *et al.*, 2017).

## *Ageratum conyzoides* (L.) L.

*Ageratum conyzoides* (L.) L. (Figure 10.10) is a noxious pan-tropical species but is native to tropical South America.

### Traditional Uses

The Iban and Malay of Sarawak mixed the plant with a little water and use the extract as eye drops for sore eyes.

### Biological Activity

Crude extracts of methanol, 40–60% petroleum ether, dichloromethane and water were screened against typed cultures of *S. aureus* NCTC 6571, Methicillin Resistant *S. aureus* NCTC 12493 and clinical isolates of resistant strains of *S. aureus, P. aeruginosa* and *Escherichia coli* (Dayie *et al.*, 2007). The methanolic extract inhibited the growth of all the strains of *S. aureus* with a zone size ranging from 26 to 28 mm in diameter. However, it was weakly active against *E. coli* and had no inhibitory activity at all against *P. aeruginosa* (Dayie *et al.*, 2007). Methanolic extract of *A. conoizydes* exhibited activity against echoviruses, a serotype of enteroviruses (Ogbole *et al.*, 2018).

**FIGURE 10.10** *Ageratum conyzoides (Photo credit: S. Teo).*

Screening of 10 African medicinal plants from Ghana showed that the hexane extract of *A. conoizydes* has a strong activity against both *Aspergillus fumigatus* as well as anti-candidal activity (Hoffman *et al.*, 2004).

Studies on the hydroalcoholic extract of *A. conyzoides* leaves were conducted to determine the anti-inflammatory effect on subacute (cotton pellet-induced granuloma) and chronic (formaldehyde-induced arthritis) models of inflammation in rats, while the absence or presence of toxicity by prolonged use of the hydroalcoholic extract was also evaluated through biochemical and hematological analysis of rats' blood samples using daily oral doses of 250 or 500 mg/kg body weight over 90 days (Moura *et al.*, 2004). The results indicated that the hydroalcoholic extract of *A. conyzoides* displayed anti-inflammatory activity but not with any apparent hepatotoxicity (Moura *et al.*, 2004).

### Chemistry

Essential oils obtained from flowers and stems of *A. conyzoides* through hydrodistillation showed antibacterial activity when tested against seven bacterial strains; the inhibition zones and minimum inhibitory concentration (MIC) for the strains which were sensitive to *A. conyzoides* essential oils were in the range of 6.7–12.7 mm and 64–256 μg/mL, respectively (Bi *et al.*, 2018). The essential oils demonstrated moderate activity against only Gram-positive bacteria *S. aureus* and *Enterococcus faecalis* (Bi *et al.*, 2018).

## *Hedyotis capitellata* Wall. ex G. Don

*Hedyotis capitellata* Wall. ex G. Don (Figure 10.11) is a weedy road-side herb common in tropical regions.

**FIGURE 10.11** *Hedyotis capitellata* (Photo credit: Shuttlestock).

## Traditional Uses

The Iban and Melanau of Sarawak squeeze the sap into infected eyes after warming the leaves over fire (Chai, 2006).

## Biological Activity

Anthraquinones are known to display antimicrobial activities such as antiviral (Barnard *et al.*, 1992), antibacterial and antifungal activities even though none of the anthraquinone isolated from *H. capitellata* had been screened against the microorganisms. A study done by Duc *et al.* (2017) showed that *H. capitellata* aqueous total extract has acute and chronic anti-inflammation activity in a dose-dependent manner.

## Chemistry

Anthraquinones had been isolated from *H. capitellata* viz., 2-hydroxymethyl-3,4-[2′-(1-hydroxy-1-methylethyl)-dihydrofurano]-8-hydroxyanthraquinone, 2-hydroxymethyl-3,4-[1′-hydroxy-2′-(1-hydroxy-1-methylethyl)-dihydrofurano]-8-hydroxyan thraquinone, 2-hydroxymethyl-3,4-[2′-1-hydroxy-1-methylethyl)-dihydrofurano]anthraquinone and 2-methyl-3,4-[2′-(1-hydroxy-1-methylethyl)-dihydrofurano] anthraquinone or capitellataquinone A (**89**), B (**90**), C and D, rubiadin (**91**), anthragallol 2-methyl ether, alizarin 1-methyl ether and digiferruginol (**92**) (Ahmad *et al.*, 2005). Some of these examples are shown in Figure 10.12.

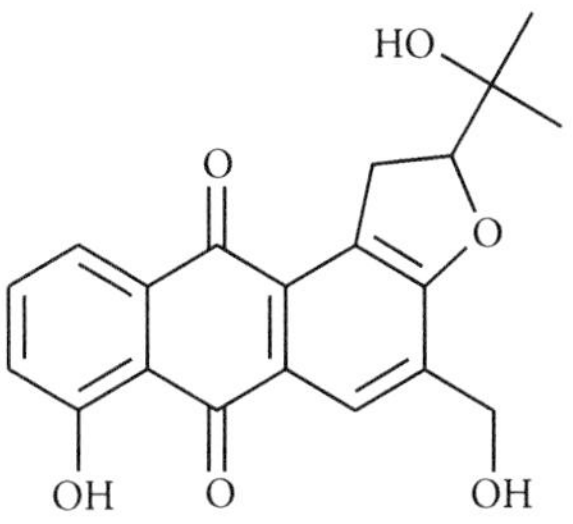

89

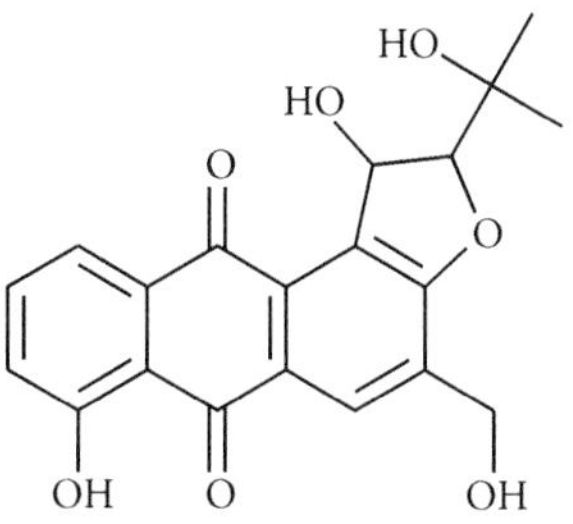

90

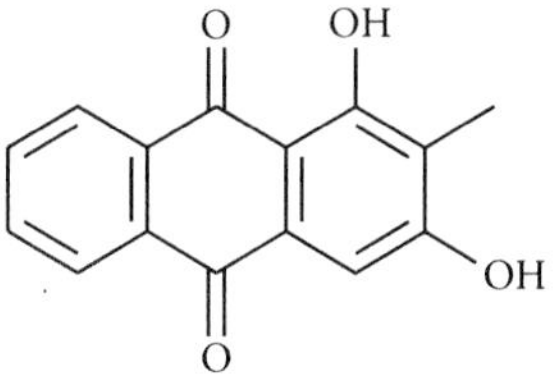

91

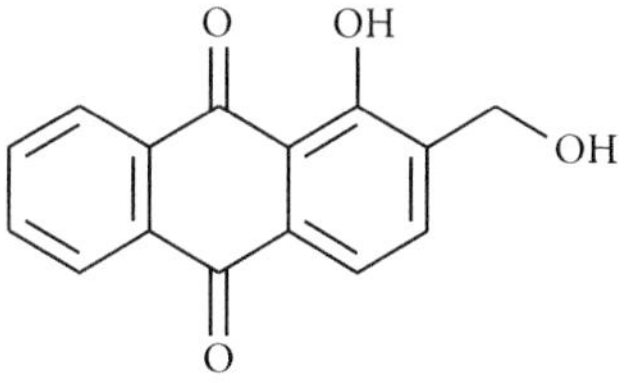

92

**FIGURE 10.12** Some of the anthraquinones isolated from *Hedyotis capitellata* – capitellataquinone A (**89**), capitellataquinone B (**90**), rubiadin (**91**) and digiferruginol (**92**).

## *MERREMIA CRASSINERVIA* OOSTSTR.

*Merremia crassinervia* Ooststr. is a creeper belonging to the Convovulaceae family that is endemic to Borneo.

### Traditional Uses

Water from outside the calyx of young fruit is squeezed into the eyes to treat eye diseases (Leaman, 1996).

### Biological Activity

There has been no work done on screening *M. crassifolia,* but other species of *Merremia* elsewhere have been screened. *Merremia emarginata* displayed anti-inflammatory activity. However, no *Merremia* species have been screened against viruses so far.

There are species of *Merremia* that possess broad-spectrum activities against Gram-positive and Gram-negative species. Pavithra *et al.* (2010) found that the methanol extract of *Merremia tridentata* exhibited significant antibacterial activity against Gram-positive and Gram-negative strains with minimum bactericidal concentration (MBC) ranging from 1.5 to 100 mg/mL. Elumalai *et al.* (2011) showed antibacterial activities against both Gram-positive and Gram-negative bacteria using the methanolic and aqueous extracts of *M. emarginata*. Methanolic extracts from *Merremia dissecta* and *Merremia aegyptia* showed antifungal activities against two fungal strains viz. *Candida albicans* and *A. niger*.

## Chemistry

Babu *et al.* (2013) isolated 4 compounds from *M. emarginata* diacetyltetritol (**93**), scopoletin (**94**), tetritol (**95**) and cynarin (**96**) (Figure 10.13). Only compounds diacetyltetritol (**93**), scopoletin (**94**) and tetritol (**95**) displayed moderate to potent anti-inflammatory activities.

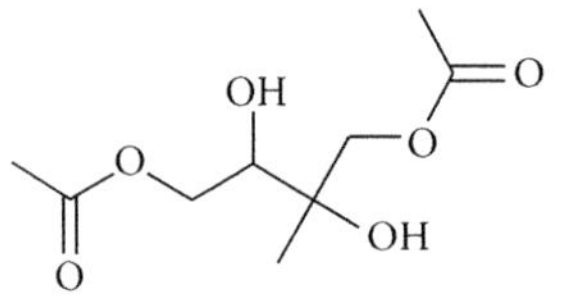

**93** 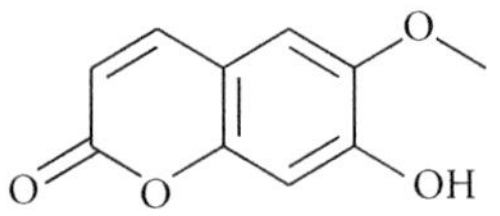 **94** 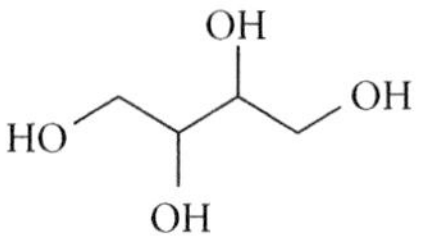 **95**

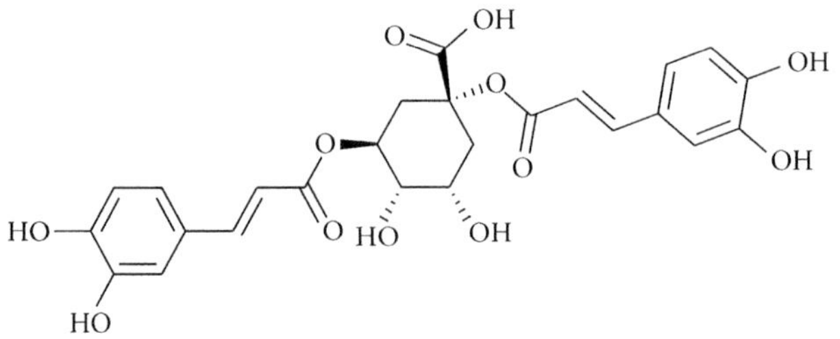

**96**

**FIGURE 10.13** Some of the compounds isolated from *Merremia emarginata* diacetyltetritol (**93**), scopoletin (**94**), tetritol (**95**) and cynarin (**96**).

# 11 Plants Used in Dental Treatment

*Stephen P. Teo*

## CONTENTS



## INTRODUCTION

Dental caries is the term used for tooth decay or cavities. It is caused by specific types of bacteria which produce acids through the conversion of sugar and carbohydrates. The acids destroy the tooth's enamel, dentine and cementum by dissolving minerals in the hard enamel that covers the tooth's crown resulting in pits which get larger over time. Globally, approximately 2.3 billion people (32% of the population) have dental caries in their permanent teeth and nearly all adults have dental caries at some point in time (Anon., 2014).

### *PALAQUIUM GUTTA* (HOOK.) BAILL.

*Palaquium gutta* (Hook.) Baill. or gutta-percha tree is a medium-sized to tall tree from the family Sapotaceae found in Southeast Asia, particularly in Malaysia and Indonesia. Gutta-percha (Figure 11.1) is a dried coagulated latex obtained from the tree; a series of cuts (concentric or v-shaped cuts) are made on the tree trunk to obtain the latex. Latex from the leaves of the tree also occasionally contributed to Gutta-percha production in the past.

#### Uses of Gutta-Percha

Gutta-percha was initially used as an insulator for cables including the trans-Atlantic cable and also for items like the golf ball. In the past, they were also used as splints for holding fractured joints and as handles of forceps, catheters, and so forth as well as to control hemorrhage in extracted socket wounds. They were also used for skin diseases by dermatologists, particularly against smallpox, erysipelas, psoriasis and eczema. Later, gutta-percha came to be associated with dentistry

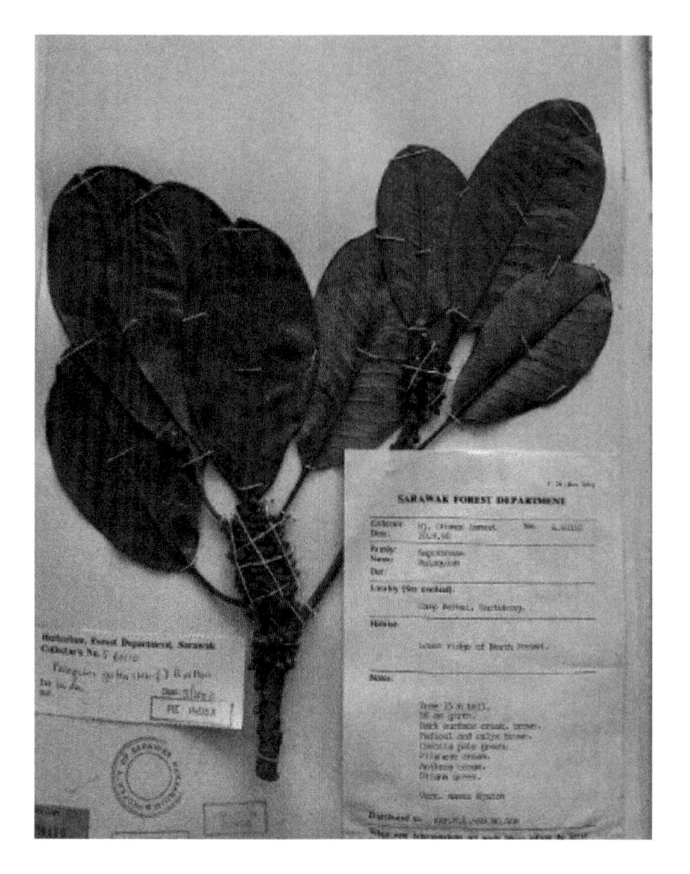


**FIGURE 11.1** *Palaquium gutta-percha* tree (Photo credit: S. Teo).

(Figure 11.2) as the thermoplastic latex is biologically inert and electrically non-conductive.

Although various methods have since been introduced since then, gutta-percha remains the main core material used for fillings in root canal treatment (Vishwanath and Rao, 2019). The many forms of gutta-percha available in the market are a reflection of technological advancement and sophistication which have helped to ease and improved efficiency in root canal sealing or obturation (Vishwanath and Rao, 2019). Barium sulfate is added to provide radiopacity for the purpose of verifying their location in dental X-ray images, whilst zinc oxide is added to increase plasticity and decrease brittleness; calcium hydroxide and/or chlorhexidine are also used to provide the antimicrobial properties (Vishwanath and Rao, 2019).

The rapid progress in dental materials science has led to the introduction of newer materials with improved properties that made such older materials outdated. However, gutta-percha's special properties of inertness, better sealing ability and the ability to do re-treatment in case of failure still make it an indispensable

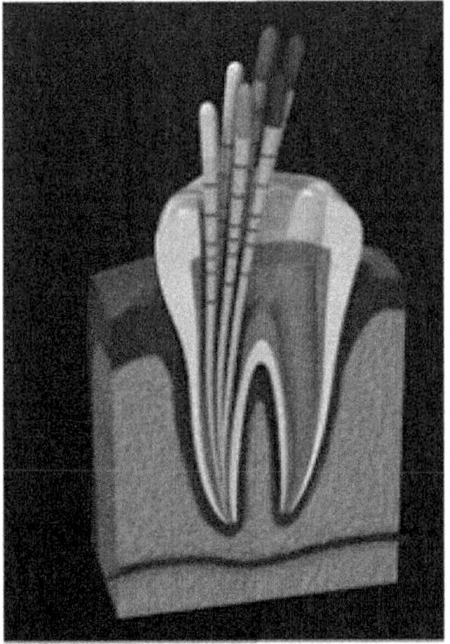

**FIGURE 11.2** Gutta-percha point used in dentistry for root canal pulp filling (Photo credit: S. Teo and shuttlestock).

obturating material. Over the years, gutta-percha has evolved and modified to suit the growing trends in endodontics.

Today, gutta-percha has also found a new use. A Swiss company imported gutta-percha from a century-old plantation in Java, Indonesia to make prosthetics – artificial limbs.

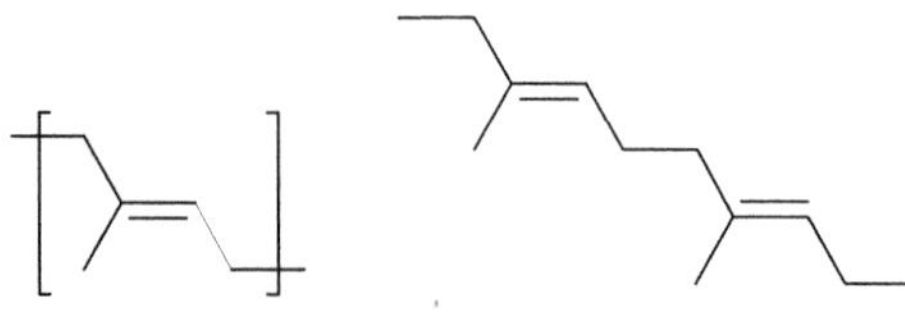

**FIGURE 11.3** Chemical structure of trans-isomer of polyispoprene (**97**).

## Chemistry and Biological Activity

Gutta-percha is a trans-isomer of polymer of isoprene or polyisopropene. Its chemical structure is 1, 4, trans-polyisoprene (**97**) (Marciano *et al.*, 1993) (Figure 11.3). The chemical structure is similar to that of natural rubber, which is a *cis*-isomer of polyisoprene. It has a number of similarities but differences in form that make its mechanical properties behave more like crystalline polymers.

The gutta-percha latex is tapped and collected in a container and boiled with water before kneading under running water. A chemical process of coagulation is done by adding alcohol and creosote mixture (20:1), ammonia, limewater or caustic soda. Using the traditional Obach's technique (Vishwanath and Rao, 2019), the resulting pulp is added to water and heated to about 75°C to obtain the gutta-percha threads, and then cooled to about 45°C. The flocculated gutta-percha or yellow gutta-percha is made up of 60% polyisoprene and 40% contaminants (resin, protein, dirt and water) and is mixed with cold industrial gasoline at below 0°C temperature, which flocculates the gutta-percha but also dissolves resins and denatures any remaining proteins. After removal of cold gasoline, de-resinated gutta threads are dissolved in warm water at 75°C and unwanted dirt particulates are allowed to precipitate. The greenish yellow residual solution is bleached and then filtered to remove any particulates, and then finally the gasoline is removed through steam distillation.

## Other Species Used Traditionally in Dental Treatment

In the past, the latex of *Dyera polyphylla* Hook. *f.* was used to treat toothache in West Kalimantan (Sudarmono, 2019) while the latexes or resins of three species from Sarawak, i.e. *Agrostistachys borneensis* Becc., *Arachnis flos-aeris* Rcbh.f. and *Bromheadia finlaysonianum* (Lindl.) Miq., were used by the Iban for treating dental caries (Christensen, 2002) (Figure 11.4).

**FIGURE 11.4** Plants (latex) used in the treatment of dental carries in Borneo. *Dyera polyphylla* (a), *Agostistachys borneensis* (b), *Arachnis flos-aeris* (c) and *Bromheadia finlaysonianum* (d).

# 12 Poisonous Plants Used as Blowpipe Dart

*Stephen P. Teo*

## CONTENTS



## INTRODUCTION

In Borneo, hunting used to be an important activity as it provided an important source of protein for the forest dwellers. Some plants that have been employed in the hunting activity as poisons, particularly in poison darts.

### *ANTIARIS TOXICARIA* LESCH.

*Antiaris toxicaria* Lesch (Figure 12.1) or ipoh belongs to a monotypic genus and is distributed throughout the Old World tropics, from West Africa to Madagascar, and in Sri Lanka, India, southern China, throughout Southeast Asia, the Pacific (east to Fiji and Tonga) and northern Australia.

#### Traditional Use

Latex from *A. toxicaria* is a notorious poison, used particularly for darts and blow-darts by all indigenous ethnic groups in Borneo in the past and is now increasingly rare and mainly restricted to the nomadic tribes (Christensen, 2002). Various traditional preparations and mixtures of "ipoh poison" are used. In general, the latex from the root-bark or bark is used alone or mixed with other poisonous and non-poisonous ingredients such as bark or roots of various species of *Strychnos*, *Derris elliptica, Hoya conorarium*, snake venoms, *Trigobalanops malayanus* and other species (Christensen, 2002). The mixture is heated over a fire and or sun dried to obtain a thick paste. The dart and arrow points are dipped directly into the mixture and then dried over fire. The mixture can be stored for later use, and the duration over which the poison retains its potency can be

**FIGURE 12.1** *Antiaris toxicaria* young leaves (left) and tree trunk with beads of poisonous latex dripping after scratching the bark (Photo credit: S. Teo).

variable, depending on mixture and method of preparation used, and can last up to 7 years (Christesen, 2002).

### Chemistry and Biological Activity

Cardenolide glycosides which are highly toxic due to their heart arresting properties that lead to death have been isolated from *A. toxicaria*. Among them are alpha- (**98**) and beta-antiarin (**99**) and toxicanoside A (**100**) (Figure 12.2). Cardenolide glycosides and coumarins isolated from *A. toxicaria* showed potent cytotoxic activities (Shi *et al.*, 2014; Kuo *et al.*, 2014).

## *Strychnos* spp.

*Strychnos* belongs to the Loganiaceae family and there are various species used for their poisons.

### Traditional Use

Roots and leaves of *Strychnos* are also used on a blowpipe dart together with the latex of *A. toxicaria*.

### Chemistry and Biological Activity

The Old World *Strychnos* species contain mainly the terpene indole alkaloids that belong to the strychnine group (**101**) (Figure 12.3), while those species from the New World contain the alkaloids from the curarine group. But other indole alkaloids had also been isolated in the old world *Strychnos* such as brucine.

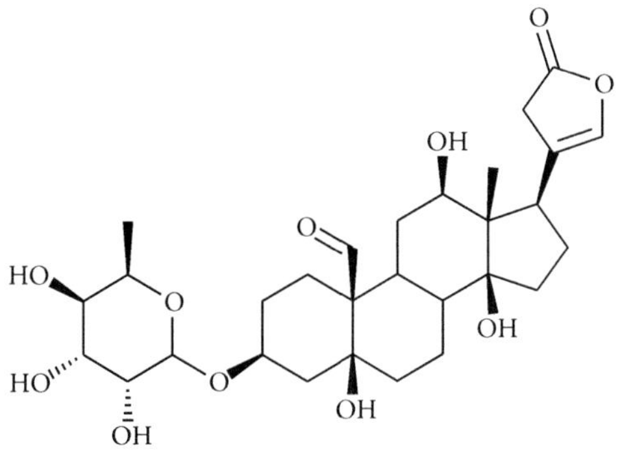

98

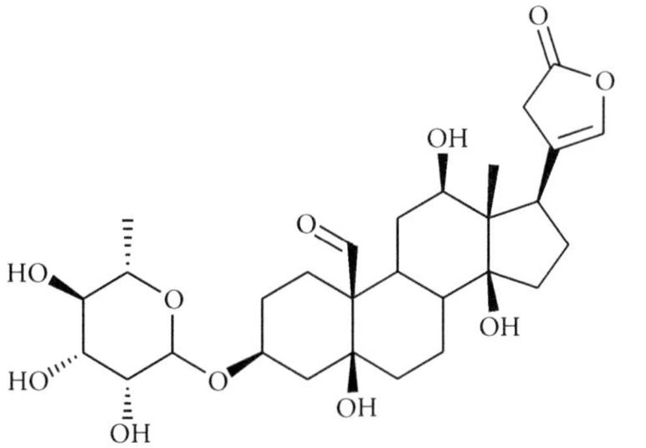

99

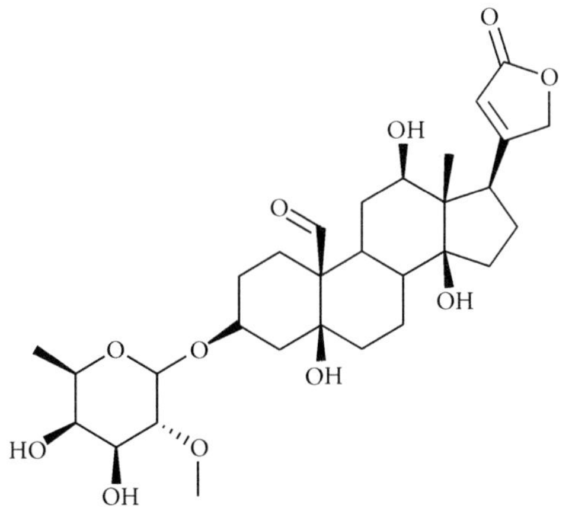

100

**FIGURE 12.2** Chemical structures of alpha- and beta-antiarin (**98** and **99**) as well as toxicarioside A (**100**).

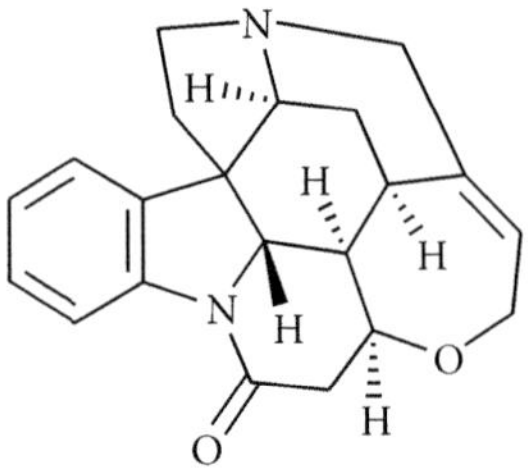

101

FIGURE 12.3 Chemical structure of strychnine (**101**).

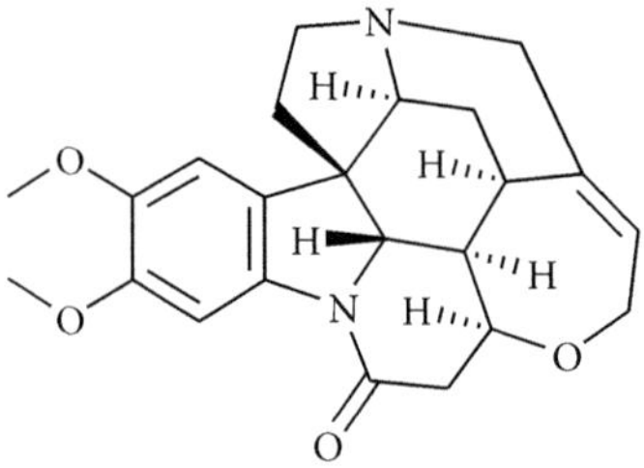

102

FIGURE 12.4 Chemical structure of brucine (**102**).

Strychnine is a neurotoxin and an antagonist of glycine and acetylcholine receptors. It primarily affects the motor nerve fibers in the spinal cord which control muscle contraction. Strychnine as an antagonist of glycine; it binds noncovalently to the same receptor, preventing the inhibitory effects of glycine on the postsynaptic neuron. Therefore, action potentials are triggered with lower levels of excitatory neurotransmitters. When the inhibitory signals are prevented, the motor neurons are more easily activated and the victim will have spastic muscle contractions, resulting in death by asphyxiation.

Brucine (**102**) (Figure 12.4) is a monoterpenoid indole alkaloid and an organic heteroheptacyclic compound. It is a white crystalline solid and is a neurotoxin and can be toxic through inhalation (vapour, dust, etc.) and ingestion.

# 13 Plants Used to Treat Arthritis

*Meekiong Kalu and Mohd Razip Asaruddin*

## CONTENTS



## INTRODUCTION

Arthritis is a common reason for absenteeism from work and can result in a decreased quality of life (OrthoInfo, 2007). According to the National Institute of Arthritis and

Musculoskeletal and Skin Diseases (NIAMS, 2014a), arthritis is a term often used to mean any disorder that affects joints; symptoms generally include joint pain and stiffness. Medically, there are about 100 forms of diseases that represent arthritis (Athanasious *et al.*, 2013). The most common forms are osteoarthritis and rheumatoid arthritis such as gout, lupus, fibromyalgia and septic arthritis (NIAMS, 2014b). The arthritic diseases affect about 15–20% of the population and become more common with age.

Since time immemorial, the indigenous people of Borneo have used various natural remedies such as herbs to treat effectively any form of arthritic diseases with few significant side effects. However, research in the field of natural remedies remains restricted. Scientific studies often use animal models or cell lines to test plant compounds and very few clinical trials exist. In this write-up, several plant species are discussed that are commonly used by the local peoples in Borneo for any form of arthritis diseases. However, many of them are not yet scientifically proven.

## *Aloe vera* (L.) Burm. *f.*

*Aloe vera* (L.) Burm. *f.* (Figure 13.1) is a succulent plant species of the family Asphodelaceae (previously placed under the family Liliaceae). It is an evergreen

**FIGURE 13.1** *Aloe vera* (Photo credit: K. Meekiong).

perennial that originated from the Arabian Peninsula, but it is now cultivated for agricultural and medicinal uses in many countries. This species is also used as an ornamental and indoor potted plant. The generic name of the plant is derived from the Arabic word, alloeh, which means bitter, referring to the taste of the sap or fluid found in the plant, while vera means "true" or "genuine". Known with many vernacular names, such as a plant of immortality, first-aid plant, miracle plant, crocodile tongue, luhui, jadam and lidahbuaya, *A. vera* can adapt to a wide range of soil and weather conditions in tropical and subtropical environments (Siti Fauziah, 2013).

## Description

A perennial herb growing to a height of about 50–70 cm tall (sometimes up to 1 m) with a stout or very short stems. The leaves are thick and fleshy, light green to grey-green, lanceolate shaped with a sharp tip and prickly spines along the margin (serrated). The leaves are arranged in a rosette. The flowers are produced on a spike up to 80–100 cm tall, and are yellow, tubular-shaped. They spread by offshoots.

## Traditional Uses

*A. vera* is often planted for commercial purposes. It is found in many consumer products including beverages, skin lotions, cosmetics, or ointments for minor burns and sunburns. Traditionally, *A. vera* is widely used by the folks to treat various skin problem or diseases, ulcer, dysentery, fever, diabetic, arthritis and headache. *A. vera* juice is applied to the abdomen of women after childbirth to give a soothing comfort, to reduce wrinkles and regain its original form. Juice from the leaves of *A. vera* also is used in treating wounds or burns and to relieve symptoms of certain skin disorders, such as rashes, psoriasis and acne.

## Chemical Constituents

*A. vera* leaves contain phytochemicals such as acetylated mannans, polymannans, C-glycosides, anthrones (**103**), anthraquinone (**104**), emodin (**19**) (Figure 13.2) and various lectins. The possible bioactivity, however, remains unstudied.

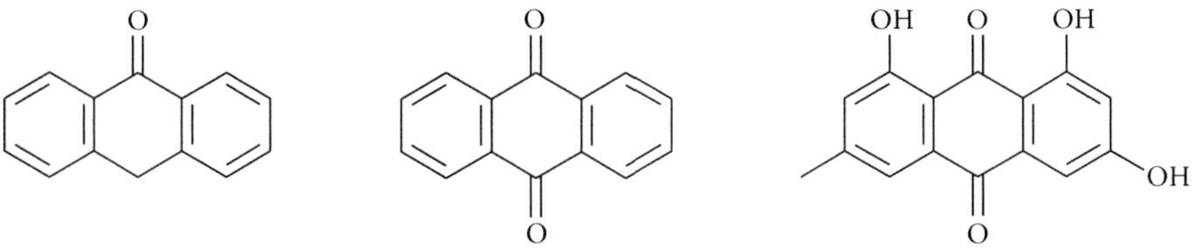

**103** **104** **19**

**FIGURE 13.2** Various anthraquinones from *Aloe vera* – anthrone (**103**), anthraquinone (**104**) and emodin (**19**).

**FIGURE 13.3** *Alpinia galanga* (Photo credit: K. Meekiong).

## *Alpinia galanga* (L.) Willd.

*Alpinia galanga* (L.) Willd. (Figure 13.3) is an herbaceous plant in the ginger family of Zingiberaceae. Known as galangal or greater galangal, this species is widely cultivated in many countries for its fragrant and spicy rhizome, rather than its dense inflorescence of small greenish-white flowers (Skornickova and Gallick, 2010). In Borneo, it is called Langkuas, and is commonly planted in the backyard gardens or orchards.

### Description

These are rhizomatous plants that form a dense clump, up to 2–2.5 m tall; have conspicuously branched rhizomes, 2–4 cm diameter, light red or pale yellow; and have a pungent smell and strong taste. The leaves are long and narrow, oblong to oblanceolate, 50–60 × 7–10 cm, glossy, nearly hairless. The inflorescence terminal of the leafy shoot appears a large cluster, spike-like or racemose, 10–35 × 4–7.5 cm, yellow-white flowers, fragrant. Fruits are rounded or ellipsoid capsule

(1–2 cm diameter), contain 2–4 seeds, yellow-orange to dark red and turn black at maturity.

### Traditional Uses

The rhizome of *A. galanga* is used most often in cooking. The fresh rhizome, flowers and young shoots are eaten raw as *ulam* (raw vegetable). The native peoples in Sarawak regularly use the rhizome of *A. galanga* for treatment of ailments including indigestion, diarrhoea, stomachache, flatulence and skin problems.

### Chemical Constituents

The rhizome of *A. galanga* contains the flavanol galangin and essential oil known as galangol. Chouni and Paul (2018) listed several active terpenoid compounds from *A. galanga*, such as *p*-hydroxycinnamaldehyde (**105**) (Figure 13.4), β-pinene, β-sitosteroldiglucoside (Ag-7), 1′-acetoxychavicol acetate (galangal acetate), α-bergamotene and so forth.

## *Curcuma longa* L.

*Curcuma longa* L. (Figure 13.5) is a flowering plant of the ginger family, Zingiberaceae. Popularly known as turmeric, this herbaceous plant is native to the Indian subcontinent and probably from Southeast Asia (Samy *et al.*, 2014). In Malaysia and Indonesia, its vernacular name is known as kunyit, temukunyit, kunyir or temukuning.

### Description

The "kunyit" plant is a perennial herb with a stout or short stem and light green tufted leaves growing up to 1 m tall. The rhizome is thick and ringed with the bases of old leaves. The leaves are large, arranged alternately in two rows, with the leaf sheaths forming a false stem-liked (called as pseudostem). Leaf blades are usually oblong, 70–120 cm (sometimes up to 250 cm) long, dark green on the upper surface, pale green beneath. Inflorescence is from the rootstock or rhizome, upright, up to 30 cm long (or longer). Flowers are yellow-white or green, sometimes tinged reddish-purple and the upper ends are tapered. They are sterile and do not produce viable seeds.

### Traditional Uses

Almost every part of *C. longa* is used by the local people of Borneo. The young rhizomes, young shoots and flowers are eaten as *ulam* (raw vegetable). The rhizomes are used fresh or dried, commonly used as a coloring and flavoring agent,

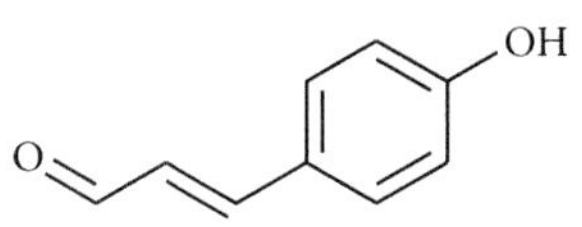

OH

O

105

**FIGURE 13.4** *p*-Hydroxycinnamaldehyde (**105**).

**FIGURE 13.5** *Cucurma longa* (Photo credit: K. Meekiong).

especially curries. According to Nelson *et al.* (2017), the turmeric powder is about 60–70% carbohydrates, 6–13% water, 6–8% protein, 5–10% fat, 3–7% dietary minerals, 3–7% essential oils, 2–7% dietary fiber and 1–6% curcuminoids. The Malays and the Melanau peoples in Sarawak use the rhizome as the main ingredient for *jamu* (beautification herb) as well as to treat joint pains.

## Chemical Constituents

Phytochemicals known from turmeric are mainly diarylheptanoids including curcurminoids such as curcumin, demethoxycurcumin and bisdemethoxycurcumin. Essential oils such as turmerone (**106**), zingiberene (**107**), curcumin (**108**) and germacrone (**109**) (Figure 13.6) are major constituents present in *C. longa* roots.

**106** **107**

**108** **109**

**FIGURE 13.6** Major constituents of essential oil in *C. longa* roots – turmerone (**106**), zingiberene (**107**), curcumin (**108**) and germacrone (**109**).

## *HIBISCUS TILIACEUS* L.

*Hibiscus tiliaceus* L. (Figure 13.7) is a species of flowering plant in the mallow family, Malvaceae, is native to tropical Asia and has become naturalized in parts of America and South America. The plant has different vernacular names, such as sea hibiscus, cottonwood, rosella, kurrajong, sea rosemallow, balibago, malabago (Philippines), waru (Javanese), barulaut (Malay), hau (Hawaiian), fau (Samoan), purau (Tahitian) and vau tree.

### Description

It is a large, stout, open-branched shrub or small tree with spreading branches, reaching a height of 4–10 m, with a trunk up to 15 cm in diameter. Leaves are heart-shaped to almost circular, 3–15 cm long. Inflorescences are cymose. Flowers are large, showy, yellow with dark maroon or blackish center, deepening to orange or apricot as they mature. Flowers fade after 1 day and turn orange-red before they are shed.

**FIGURE 13.7** *Hibiscus tiliaceus* (Photo credit: K. Meekiong).

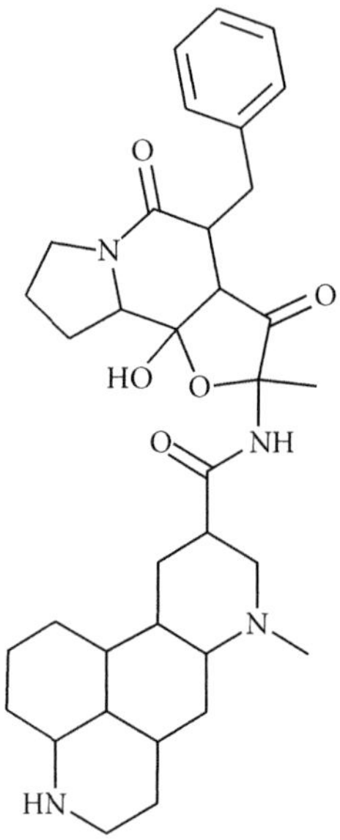

110

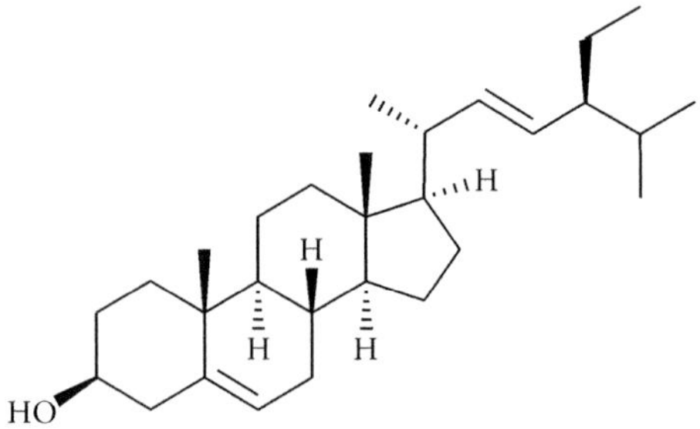

111

**FIGURE 13.8** Chemical constituents of *Hibiscus tiliaceus* – dihydroxyergostenedione (**110**) and stigmasterol (**111**).

## Traditional Uses

*H. tiliaceus* is widely used in Asian nations in many ways. They are occasionally found as ornamental landscape trees along the road or as the subject for the art of bonsai (especially in Taiwan). The leaves are used for wrapping food or for fermenting tempeh. The wood of *H. tiliaceus* has been used in a variety of applications such as light construction, firewood and carvings. The fibers extracted from the bark or stem have traditionally been used in rope making. The young leafy shoots are eaten as *ulam*. The bark and roots are used as traditional medicine by boiling in water and taken as a drink to cool a fever. The local peoples in coastal areas of Sarawak use the crushed leaves of *H. tiliaceus* and mix with other selected plants (e.g. *Ardisia elliptica, Zingiber officinale, Saccharum officinale*) to treat broken bones or joint pains.

## Chemical Constituents

Little *et al.* (1989) noted that leaves of *H. tiliaceus* displayed strong free radical scavenging activity and contain tyrosinase activity, whilst Melecchi *et al.* (2002) reported many compounds have been found in the extraction of *H. tiliaceus* flower that include octadecenoic acid, dihydroxyergostenedione (**110**), dihydrobenzofuran, gynmolutone and stigmasterol (**111**) (Figure 13.8).

## *MORINDA CITRIFOLIA* L.

*Morinda citrifolia* L. (Figure 13.9) is a flowering plant of the coffee family, Rubiaceae. It is native to Southeast Asia and Australasia and is cultivated throughout the tropics as an ornamental or for medicinal purposes and has become

**FIGURE 13.9** *Morinda citrifolia* – leaves (top left); ripe fruit; (right) inflorescence of *Morinda citrifolia*. (right) (Photo credit: K. Meekiong).

naturalized. Known by many names across the globe, the common English names are Great morinda, Indian mulberry, noni, cheese fruit and beach mulberry. In Borneo, it is called mengkudu or engkudu.

## Description

It is a small evergreen tree or shrub, up to 10 m in height, with spreading branches, and bushy-like. Leaves are large, membranous, 18–50 × 6–30 cm, glabrous, pinnately veined and glossy. Flowers are many, 75–90 in ovoid to globose heads. The corolla is white, 5-lobed. It produces syncarp fruits, yellowish-white, fleshy, 6–12 cm long (sometimes longer), 4–5 cm in diameter, soft and fetid when ripe.

## Traditional Uses

Despite its strong smell and bitter taste, the fruits are eaten during famine or as staple food, either raw or cooked. The local people consume the fruit raw with salt or cook it with curry. The ripe fruits are blended to make juice and a drink to lower the sugar or glucose content in the blood. The fruits and leaves of *M. citrifolia* also are used to cure fever and arthritic diseases. The roots are febrifuge, tonic and antiseptic and used to treat stiffness and tetanus and have been proven to combat arterial tension. An infusion of the root is used in treating urinary disorders and the bark is used in a treatment to aid childbirth (World Health Organization, 1998).

## Chemical Constituents and Prospects

According to Inada *et al.* (2017) and Almeida *et al.* (2019), *M. citrifolia* has high nutritional value and almost 200 phytochemical compounds with bioactive properties have already been identified and isolated from different parts of the plant. Su *et al.* (2005) reported 17 common compounds found in the fruits of *M. citrifolia.* Amongst them are americanin A, narcissoside (**112**), asperuloside, borreriagenin, citrifolinin B epimer a, citrifolinin B epimer b and cytidine. Kamiya *et al.* (2010) reported active compounds, damnacantha (**113**), nordamnacanthal, dihydroxy-2-methoxymethylanthraquinone and morindone (**114**) (Figure 13.10)

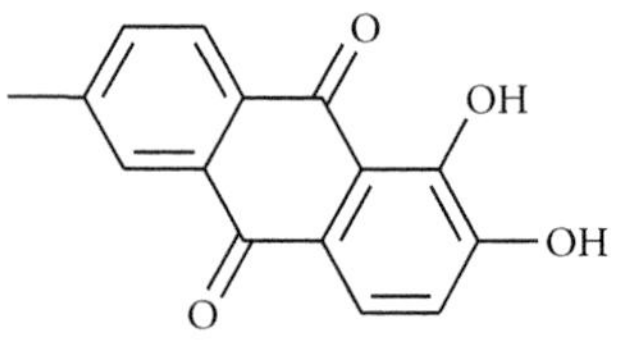

**112**

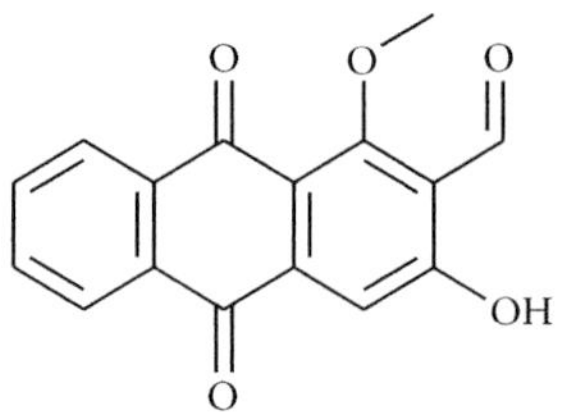

**113**

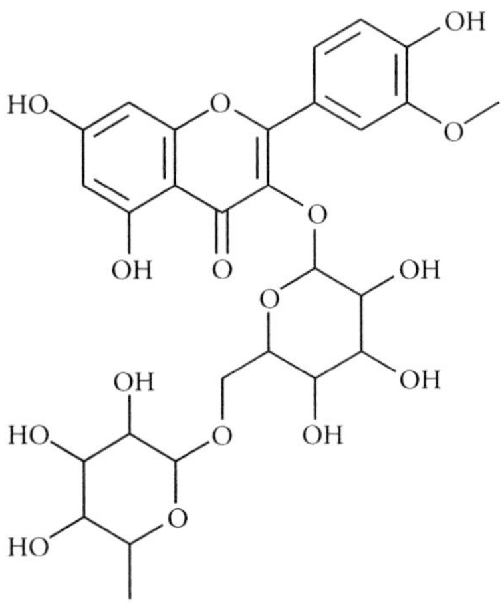

**114**

**FIGURE 13.10** Major constituents of *Morinda citrifolia* – narcissoside (**112**), damnacantha (**113**), and moridone (**114**).

that were extracted from the roots of *M. citrifolia* and showed positive results against cancer.

## *Mimosa pudica* L.

*Mimosa pudica* L. (Figure 13.11) is a creeping perennial flowering plan of the legume family, Fabaceae. The epithet name, pudica, is a Latin word meaning shy, bashful or shrinking, referring to the compound leaves that fold inward and droop when touched or shaken, and recognized as 'sleeping movement' also known as nyctinasty movement. It is known with many vernacular names, viz. sensitive plant, sleepy plant, touch-me-not, shy plant, zombie plant and semalu or malu-malu in Malay. This species is native to the New World (South and Central America) but is now a cosmopolitan weed. It is a pioneer plant species, chiefly found on soils with low nutrient concentrations and rapidly spread over the area.

### Description

The stem is slender, branching and sparsely to densely covered with hairs and prickly-like structures; erect in young plants and becomes creeping or trailing with age; growing to a length up to 1.5 m long trailing. Leaves are compound, 10–28 leaflets per pinna. Its inflorescence with flower head arises from the leaf axils. Flowers are pink or purple. The fruits consist of cluster of 2–10 pods, 1–3 cm long; the pods break into 2–5 segments contain pale brown seeds.

### Traditional Uses

The Malay people used the boiled water of *M. pudica* to treat fever, insomnia, stress and rheumatism. The Iban from the rural areas of Sarawak the boiled water of *M. pudica* roots was drunk for treatment of asthma, ulcer and malaria. The dried plants of *M. pudica* are mixed with other plants (e.g. *A. galanga, Z. officinale, Glochidion* sp., *S. officinale, A. elliptica*) as a poultice for broken bone treatment.

### Chemical Constituents

Genest (2008) and Parasuraman *et al.* (2019) reported that *M. pudica* contains various compounds, including alkaloids, flavonoid C-glycosides, sterols (**115**), terpenoids, tannins, saponin and fatty acids. Additionally, extracts of *M. pudica* have been shown to contain crocetin-dimethylester (**116**), tubulin and green-yellow fatty oils. A new class of phytohormoneturgorines, which are derivatives of gallic acid, gallic acid 4-*O*-(β-D-glucopyranosyl-6′- sulfate) (**117**), have been discovered within the plant (Figure 13.12) (Azmi, 2011).

## *Zingiber officinale* Roscoe

*Zingiber officinale* Roscoe (Figure 13.13) is an herbaceous plant of the flowering family Zingiberaceae. Known as the true ginger, *Z. officinale* has been grown in Asia since ancient time. Although the origin is still vastly debated, many believe that this species originated in India and was taken to Europe and East Africa by the

**FIGURE 13.11** Creeping plants of *Mimosa pudica* (Photo credit: K. Meekiong).

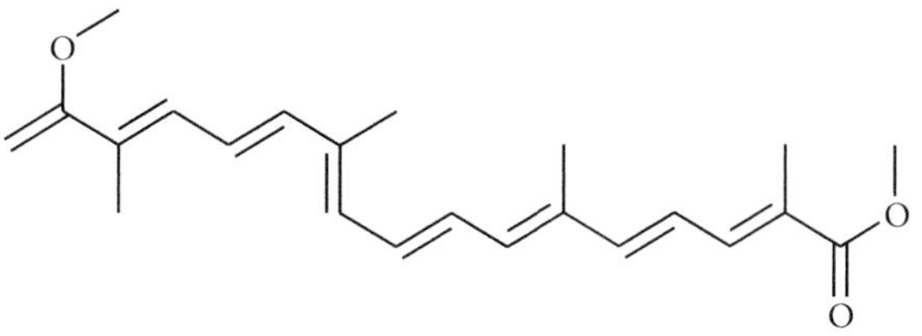

**115**

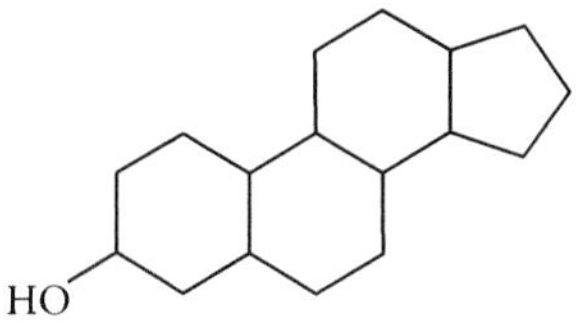

**116** 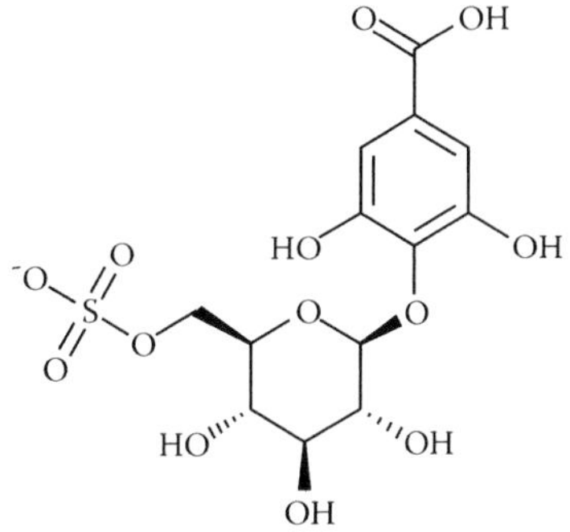 **117**

**FIGURE 13.12** Major constituents of *Mimosa pudica* – sterol (**115**), crocetin-dimethylester (**116**) and gallic acid 4-*O*-(β-D-glucopyranosyl-6′-sulfate) (**117**).

Arab traders as a precious item of trade. The generic name might be derived from the Sanskrit word, meaning "body with horn", referring to the shape of the root, and the epithet name *officinale* is a Latin word, mean medicinal. In Malaysia and Indonesia, it is popularly known as halia.

## Description

True ginger is an aromatic small herb with underground rhizome, erect, up to 75 cm tall. The leaf sheaths are overlapping, rolled and clasping to develop a false stem or pseudostem. Leaves are simple, sheathing at the base, linear-lanceolate, 15–20 × 3–4 cm, glabrous, green or dark green above and pale green beneath. Its inflorescence from rootstock or rhizome is erect, with a spike on a distinct scape. Flowers are densely arranged, irregular, yellow or yellowish-green with purplish spots. The fruit is an oblong capsule, with many seeds.

## Traditional Uses

True ginger is a fragrant kitchen spice, regularly used as an ingredient in many dishes. The roots can be made into candy or ginger wine or can be steeped in boiling water to make ginger herb tea. The Malay and the Melanau peoples in the central part of Sarawak commonly use ginger as the main ingredients to make poultices to treat arthritis. The rhizomes of *ginger* are boiled and taken as a drink to relieve wind in the body.

**FIGURE 13.13** *Zingiber officinale* (Photo credit: K. Meekiong).

## Chemical Constituents

Many studies have been conducted to identify and isolate the active compounds from different parts of *Z. officinale*. Among the studies were those by Liu *et al.* (2019), Bhattarai *et al.* (2018), Sharma *et al.* (2016), Pilerood and Prakash (2011), Charles *et al.* (2000), Connell and McLachlan (1972) and so forth. There are more than 400 chemical constituents isolated and identified from the *Z. officinale* including volatile oils e.g. zingerol (**118**), borneol (**119**), cineole, citronellol, nerol, terpinolene, gingerols (e.g. gingerol (**120**), acetoxy-4-gingerol, 10-gingerdione, paradol, shogaol, zingerine, zingerone and so forth and diarylheptanoid (e.g. 1,7-bis (4′hydroxy-3′-methoxyphenyl)-4-heptene-3-one, 1,7-bis (4′-hydroxy-3′-methoxyphenyl)3,5-heptadione and so forth). Some examples of these compound structures are shown in Figure 13.14.

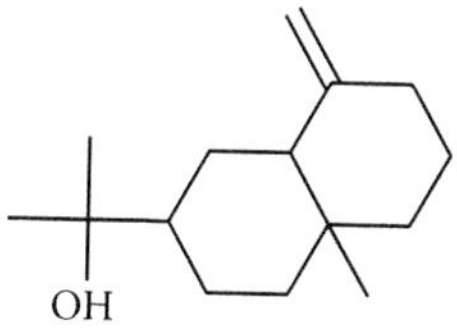

**118**

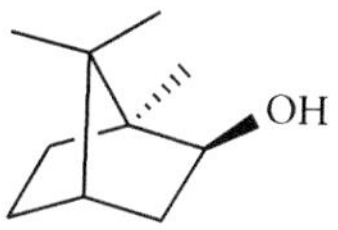

**119**

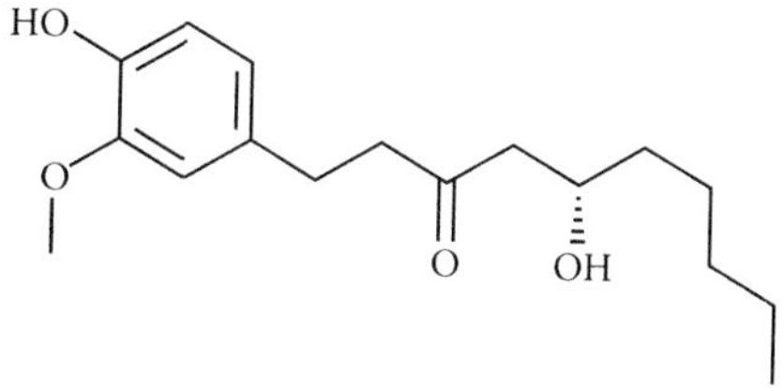

**120**

**FIGURE 13.14** Chemical constituents of *Zingiber officinale* – zingiberol (**118**), borneol (**119**) and gingerol (**120**).

# 14 Plants Used for Wound Healing

*Tukirin Pratomidhardjo*

## CONTENTS

## INTRODUCTION

Wounds are physical injuries that result in an opening or breaking of the skin whilst proper healing of wounds is necessary for the restoration and functional status of the skin (Kumari *et al.*, 2016). Wound healing is a normal biological process that the body carries out to repair wounds as a response to tissue injury and it is interdependent on the cellular and biochemical stages (Farahpour, 2019). Delayed wound healing increases the chance of microbial infection or interference in the healing process, while an improved wound healing process can shorten the time required for healing (Farahpour, 2019).

The wound healing process is achieved through four precisely programmed and highly overlapping phases: (1) Hemostasis, (2) Defensive/Inflammatory, (3) Proliferative, and (4) Maturation/Remodeling (Maybard, 2020; Guo and DiPietro, 2010). In the first phase, the body activates the blood clotting system and during this process, platelets come into contact with collagen, resulting in activation and aggregation while the thrombin enzyme initiates the development of a fibrin mesh that strengthens the platelet clumps into a stable clot (Maynard, 2020). The second phase, called the Defensive/Inflammatory Phase, concentrates on eradicating bacteria and clearing debris – basically preparing the wound bed for new tissue to form (Maynard, 2020). Next, the wound enters Phase 3, the Proliferative Phase, and the target is to fill and cover the wound which features three distinct stages: (1) filling the wound; (2) contraction of the wound margins; and (3) covering the wound or epithelialization (Maynard, 2020). Finally, during the Maturation or Remodeling Phase, the new tissue slowly gains strength and flexibility. Through the reorganization of collagen fibers, the tissue remodels and matures and there is an overall increase in tensile strength (though maximum strength is limited to 80% of the pre-injured strength) (Maynard, 2020).

All four phases must occur in the proper sequence and time frame (Guo and DiPietro, 2010). Many factors can interfere with one or more phases of this process, thus causing improper or impaired wound healing (Guo and DiPietro, 2010).

Medicinal plants can promote wound healing effects by different mechanisms such as by modulation in wound healing, promoting blood clotting, fighting against infections and accelerating wound healing including improving collagen deposition, increasing fibroblasts and fibrocytes, and so forth (Farahpour, 2019). The availability of synthetic drugs capable of stimulating the process of wound repair is still limited, so there is increasing interest in finding herbs that have wound healing efficacy (Kumari *et al.*, 2016).

## *Nephelium lappaceum* L.

*Nephelium lappaceum* L. (Figure 14.1) also known as rambutan is a seasonal tropical fruit tree species commonly grown in Southeast Asia and belongs to the same family as lychee and longan, the Sapindaceae.

### Traditional Uses

In Sarawak, the leaves are boiled for a short time and the juice squeezed from the boiled leaves is dripped into the wound or cut (Christensen, 2002).

**FIGURE 14.1** *Nephelium lappaceum* (Photo credit: S. Teo).

### Biological Activity

Little experimental work has been done to determine the wound healing effect by leaf extract from *N. lappaceum* and majority of the work made use of the fruit peel. Subramanian *et al.* (2018) used a 2,2-diphenyl-1-picrylhydrazyl (DPPH) model of *in vitro* free radical scavenging test with *N. lappaceum* leaf extracts. The extract was also tested on wound healing activity using a mice excision model, while the anti-inflammatory effect was tested in mice using the formalin induced paw licking test model. In the DPPH test, *N. lappaceum* leaf extract at all concentrations were found to cause significant free radical inhibition ($p < 0.05$). All concentrations of *N. lappaceum* leaf extract showed significantly ($p < 0.05$) high wound contraction on the 6th day. In the animal model of inflammation, all doses of *N. lappaceum* showed a reduced amount of paw licking (Subramanian *et al.*, 2018).

### Chemistry

Compounds isolated from *N. lappaceum* were mainly from the peels and they were saponin and phenolics, especially flavonoids. The isolated compounds from the *N. lappaceum* peels were identified as ellagic acid (EA) (**121**), corilagin (**122**) and geranin (**123**) (Thitilertdecha *et al.*, 2010) (Figure 14.2). *N. lappaceum* and its flavonoid compounds have the ability to counteract destructive cell damage and re-establish normal wound healing-induced fibroplasia by TGF-β1. Besides, *N. lappaceum* may promote wound healing due to the presence of geranin that reduces or inhibits hyperglycemia (Subramanium *et al.*, 2018).

## *Morinda citrifolia* L.

*Morinda citrifolia* L. (Figure 14.3) is also called noni or Indian mulberry. It belongs to the Rubiaceae family and is a secondary forest species often found growing in open and abandoned spaces or roadside. It is a native to tropical and subtropical regions and the Pacific Islands.

### Traditional Use

In Sarawak, fresh leaves are crushed and then rubbed onto the wound (Chai, 2006).

### Biological Activity

A significant enhancement in the wound healing activity has been reported in the animals treated with the *M. citrifolia* extract compared to animals receiving the placebo control treatments. *M. citrifolia* extract improves wound healing by decreasing wound size and time for epithelialization (Lee *et al.*, 2005). Furthermore, in an animal study with an ethanolic extract and drinking water as control, it was demonstrated that on day 11, the extract-treated animals displayed a 71% reduction in the wound area with enhanced wound contraction, decreased epithelialization time, increased hydroxyproline content as compared with 57% for controls (Nayak *et al.*, 2007). There were also a significant increase in granulation tissue weight and hydroxyproline content in the dead space wounds in extract-treated animals compared with the controls ($p < 0.002$) (Nayak *et al.*, 2007). Nayak *et al.* (2009) also found that ingestion of the ethanolic extract of *M. citrifolia* leaves increased

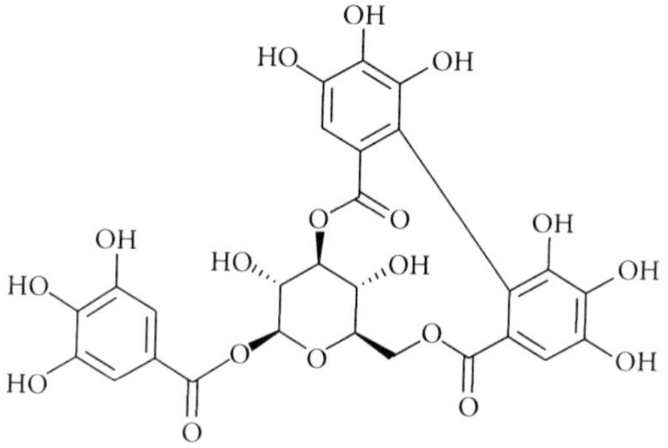

121

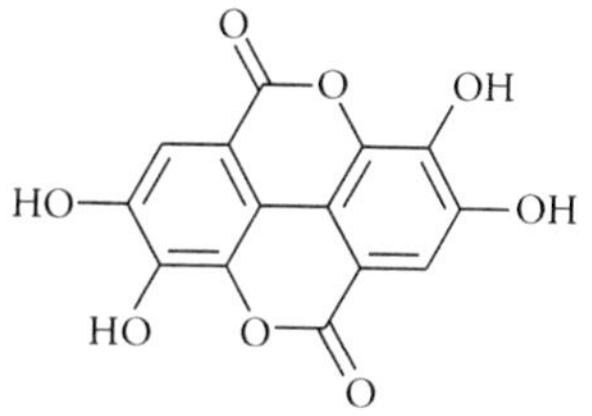

122

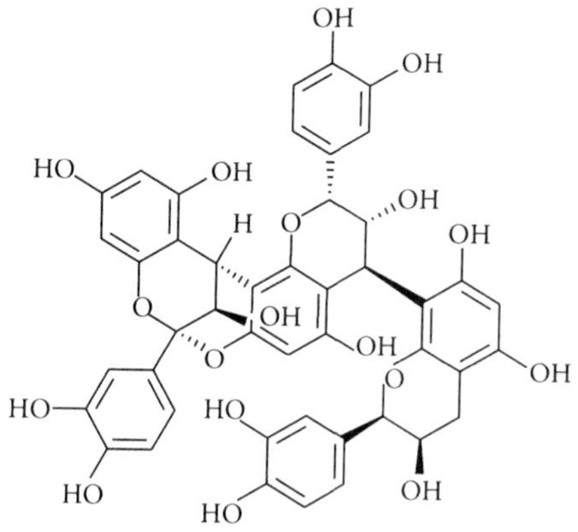

123

**FIGURE 14.2** Chemical structures of ellagic acid (**121**), corilagin (**122**) and geranin (**123**).

excision wound closure rates and promoted significantly greater levels of hydroxyproline in dead space wounds of Sprague Dawley rats.

Irfan and Desy (2018) conducted laboratory experimental study with a sample of 27 white rats divided into 3 groups (treatment with *M. citrifolia* leaf extract, positive and negative control) was carried out by measuring and observing the duration of wound healing post tooth extraction, starting from the formation of sockets after tooth extraction until the socket closure. The average healing time in the negative control group was 19 days, 9 days and 7 days for the positive control, negative control and the treatment group with *M. citrifolia* leaves extract respectively (Irfan and Desy, 2018).

Experiments had also been conducted with cream, paste or gel with *M. citrifolia* extract. In an experiment with 24 Wistar rats divided into 4 groups with incisional wounds with 4 different treatments (placebo cream, 5%, 10% and 20% *M. citrifolia*

**FIGURE 14.3** *Morinda citrifolia* (Photo credit: S. Teo).

ethanolic extract cream) and terminated on the 15th day, microscopic observation of scar tissue using Picrosirius Red (PR) staining showed that the rat groups at all concentrations of *M. citrifolia* have a higher mean percentage of collagen deposition when compared with the base cream application only (Lay *et al.*, 2019). Another experiment using 30 Dawley rats with the lower left incisor tooth extracted and divided into three groups *M. citrifolia* (gel, poviclone iodine, and control) showed that the application of *M. citrifolia* gel can accelerate the escalation of fibroblast post tooth extraction. Examination showed there was significant difference in fibroblast amounts between the group with *M. citrifolia* gel and the two other groups ($p < 0.05$) (Khoswanto, 2010).

In an investigation using ethanol-based *M. citrifolia* paste on oral mucosa wounds in Wistar rats that were divided into 2 control groups (no medication and ethanol gel with a concentration of 10% *M. citrifolia* leaf extract) and 4 treatment groups (pastes formulated in concentrations of 2.5%, 5%, 10%, and 20%) based on visual wound closure and fibroblast cell counts, results revealed that all groups treated with *M. citrifolia* leaf paste demonstrated better wound closure ($p < 0.05$) when compared to the control groups (Sabirin and Yuslianti, 2016). It was further shown that fibroblast cell counts showed little significance amongst all groups ($p = 0.143$) but the fibroblast cell counts of groups treated with *M. citrifolia* leaf paste (of all concentrations), were lower than both control groups.

Palu *et al.* (2010) demonstrated that the probable mechanisms of wound healing by *M. citrifolia* leaves were via its ligand binding to the platelet-derived growth factor (PDGF) and human adenosine $A_{2A}$ receptors as fresh *M. citrifolia* leaf juice showed significant affinity to PDGF receptors, and displayed 16% binding inhibition of the ligand binding to its receptors, while at the same concentration, it only had 7% inhibition of the ligand binding to the human adenosine $A_{2A}$ receptors. Both hexane and methanolic fractions of leaf ethanolic extract showed significant affinity to $A_{2A}$ receptors, which were dependent on concentration. However, when compared with the control, the methanolic fraction of the ethanolic extract significantly increased wound closure and reduced the half closure time in mice with a $CT_{50}$ of $5.4 \pm 0.2$ days ($p < 0.05$) (Palu *et al.*, 2010).

Numerous *M. citrifolia* products have been commercialized and sold online as well.

### Chemistry

No bioactive compounds have ever been attributed to wound healing activity of *M. citrifolia* leaves.

## *Melastomata malabathricum* L.

*Melastomata malabathricum* L. is a secondary forest species often found growing in open spaces as well as on roadsides. It is native to tropical and subtropical Asia and the Pacific Islands.

### Traditional Use

In West Kalimantan, the leaves are chewed and the poultice pasted on the wound by the Dayak Jangkang Tanjung (Sari *et al.*, 2014) or rubbed onto the wound by the Seberuang Dayak Tribe (Takoy *et al.*, 2013). In Sarawak, young leaves as well as burnt roots are pounded into powder and applied topically to stop bleeding by the Iban of Sarawak (Chai, 2006).

### Biological Activity

Wound healing effect of methanolic ointment of *M. malabathricum* extract was tested in two types of wound models in rats: (i) the excision wound model and (ii) the incision wound model. The methanolic extract ointment displayed a significant response in both models tested which is comparable with the standard drug (nitrofurazone) in terms of wound contracting ability, wound closure time, tensile strength and regeneration of tissues at the wound site (Sunilson *et al.*, 2018). The *M. malabathricum* methanolic and ethanolic extracts also showed broad spectrum anti-bacterial activity (Alwash *et al.*, 2014; Sunilson *et al.*, 2018). For the methanolic extract of *M. malabathricum*, it displayed anti-bacterial activities against the various clinical wound bacterial isolates such as *Staphylococcus aureus* and *Pseudomonas aeruginosa* with MIC values ranging from 3.0 to 8.0 mg/ml (Sunilson *et al.*, 2018).

Using ethanol-induced and indomethacin-induced ulcer models in rats, anti-ulcer effects of the ethanolic extracts at 250 and 500 mg/kg were evaluated, which demonstrated that the ethanolic extract of *M. malabathricum* exhibited significant and dose dependent anti-ulcer activity in the models used (Balamurugan *et al.*, 2013). The percentage of ulcer inhibitions by the extract at 500 mg/kg for ethanol- and indomethacin-induced ulcer was 64.3 % and 73.8% respectively (Balamurugan *et al.*, 2013). Ulcer protection in the model used by the extract was dose dependent and the ulcer inhibitory effects of the extract were comparable to omeperazole (Balamurugan *et al.*, 2013).

### Chemistry

An aqueous extract of *M. malabathricum* leaves, screened using qualitative and quantitative methods, showed the presence of high concentration of flavonoids (10.8 mg/ml) than tannins (6.2 mg/ml) in the extract (Nurdiana and Marziana, 2013). The flavonoid-rich aqueous extract was tested for wound healing using twelve Sprague Dawley rats, each with 1 $cm^2$ of excision on their back were divided into four groups and given different treatments. Group 1, was treated with an aqueous extract of *M. malabathricum* leaves, Group 2 (treated with providerm) and Group 3 (treated with acriflavin) as positive controls using conventional drugs and Group 4 – saline as negative control (Nurdiana and Marziana, 2013). *M. malabathricum* showed the highest percentage of wound contraction (93%) on the 15th day, followed by flavin (88%), Poviderm (86%) and negative control (77%). Rats with *M. malabathricum* leaf extract treatment also showed the finest scar on the wound slit with little inflammation and no microbial infection compared to other treatments (Nurdiana and Marziana, 2013).

## *Ageratum conyzoides* (L.) L.

*Ageratum conyzoides* (L.) L. (Figure 14.4) is a herbaceous plant that belongs to Asteraceae or Compositae family. It is commonly known as Billy goat weed and is found growing in some regions of Africa, Asia and tropical America. It is native to

**FIGURE 14.4** *Ageratum conyzoides*. (Photo credit: Shuttlestock).

tropical America, especially Brazil and is considered an invasive weed in many other regions.

### Traditional Uses

Leaves and shoots were mashed and applied to the wounds in Kalimantan, while the various ethnic groups in Sarawak pounded or rolled the fresh leaves into a paste to be applied to the wound (Chai, 2006).

### Biological Activity

In an animal study conducted, methanolic extract of *A. conyzoides* as opposed to saline solution used in the control group exhibited a significant increase in the percentage of wound contraction at day 10 in the experimental group compared with the control. It was further concluded that extracts of *A.conyzoides* has a better wound healing enhancing action compared with normal saline treated control (Oladejo *et al.*, 2003).

In another experiment using 200 μL of ethanolic *A. conyzoides* extract and 200 μL of 50% ethanol as control respectively on open incision wounds in rats, the *A. conyzoides* extract was found to increase cellular proliferation and collagen synthesis (Arulprakash *et al.*, 2012). Wounds treated with the extract were found to heal much faster, based on the improved rates of epithelialization and wound contraction, and on the histopathological results. Furthermore, a 40% increase in the tensile strength of the treated tissue was also observed (Arulprakash *et al.*, 2012).

A study on the hemostatic effect of methanolic leaf extract of *A. conyzoides* using albino rat as a model indicated that methanolic leaf extract of *A. conyzoides* significantly decreased ($p < 0.05$) the bleeding time, prothrombin time and clotting time respectively in a dose dependent manner. In contrast, there was a significant increase in plasma fibrinogen concentration ($p < 0.05$) (Bamidele *et al.*, 2010).

Igboasoiyi *et al.* (2007) demonstrated that the ethanolic extract of *A. conyzoides* at the dose of 500 and 1000 mg/kg administered orally and daily for a one month period revealed that it was not toxic in rats since the $LD_{50}$ value confirmed that *A. conyzoides* is safe for use.

### Chemistry

So far, no work has been done to isolate the bioactive chemical compounds that contribute to wound healing activity.

## *MUSA × PARADISIACA* L.

### Traditional Use

The stem exudates from *M. × paradisiaca* are rubbed on the wound by various Dayak tribes of West Kalimantan (Meliki *et al.*, 2013; Rufina *et al.*, 2013)

### Biological Activity

Most activities related to *M. × paradisiaca* were performed using the pulp and peel (Imam and Akter, 2011). However, there were a few studies conducted on the slimy sap of the stem. Results indicated that the *M. × paradisiaca* L. extract showed antibacterial activity against *P. aeruginosa* and *S. aureus* with the zone of inhibition of 21 mm and 19 mm respectively at concentration of 500 μg/disc (Amutha and Selvakumari, 2016).

Wistar albino rats were selected for burn wound healing activity created by using red hot steel rod from above the hind limb region with methanolic extract applied on the wound and monitored daily. The wound contraction rate was absorbed based on the histopathological examination. It was concluded that the methanolic extract of *M.* × *paradisiaca* showed greater healing activity compared to control in Wistar albino rats (Amutha and Selvakumari, 2016).

An ointment containing extract of plantain stem juice (10%) was formulated and tested for wound healing activity in rats using excision wound model. The results indicated that topical application of the formulated ointment significantly ($p$ <0.05) enhanced the rate of wound healing whilst shortening the epithelization period (Adenortey, 2018). The closure of wound area for the ointments of plantain stem juice and silver sulphadoxine were 98.9 ± 0.7 % and 100 ± 0.00%, respectively. The taken for epithelization was greatly reduced from 21.0 ± 1.4 days for the petroleum jelly-treated group to 14.6 ± 0.5 days for silver sulphadoxine-treated group and 16.8 ± 0.8 days for the extract-treated group (Adenortey, 2018).

### Chemistry

There is no chemistry work done to isolate the bioactive compounds so far.

## *Gynura procumbens* (Lour.) Merr.

*Gynura procumbens* (Lour.) Merr. (Figure 14.5) is an edible herbaceous species found growing wild or cultivated as vegetable or for medicinal purposes.

**FIGURE 14.5** *Gynura procumbens* r. (Photo credit: Shuttlestock).

## Traditional Use

The Penan of Sarawak make use of the ash from the leaves for wound healing.

## Biological Activity

In a study on the topical application of ethanolic leaf extract of *G. procumbens* on the rate of wound healing closure and histology of wound area and from macroscopic observation, wound with topical application of *G. procumbens* leaf extract and Intrasite gel significantly healed earlier than those treated with vehicle (Zahra *et al.*, 2011) while the histological analysis of wounded area of 6 animals from each group which were sacrificed on the 14th day after wounding showed that wounds dressed with leaf extract showed comparatively less scar width at wound closure and granulation tissue had fewer inflammatory cells and more collagen with angiogenesis compared to wounds applied with vehicle (Zahra *et al.*, 2011).

In an examination on the effects of *G. procumbens* gel treatment on wound healing in streptozotocin-induced diabetic mice, Sutthammikorn *et al.* (2018) found that *G. procumbens* treatment significantly promoted wound healing faster than the standard wound healing drug in diabetic patient (solcoseryl jelly) by promoting angiogenesis around the wound area, and significantly increased the skin expression of angiogenin, endothelial growth factor, fibroblast growth factor, transforming growth factor and vascular endothelial growth factor in both normal and diabetic mice. Besides, *G. procumbens* elevated the expression of numerous growth factors in human fibroblasts, keratinocytes, endothelial cells and mast cells apart from promoting keratinocyte and fibroblast proliferation and fibroblast, keratinocyte and mast cell migration (Sutthammikorn *et al.*, 2018).

Rosidah *et al.* (2009) noted that oral administration of the methanol extract for a period of 13 weeks from *G. procumbens* leaves at 1000–5000 mg/kg did not produce mortality or significant changes in the general behavior, body weight, or organ gross appearance of rats (Rosidah *et al.*, 2009). There were no significant differences in the general condition, growth, organ weights, hematological parameters, clinical chemistry values, or gross and microscopic appearance of the organs from the treatment groups as compared to the control group. Therefore, the no-observed-adverse-effect-level NOAEL for the *G. procumbens* methanol extract is 500 mg/(kg day) administered orally for 13 weeks (Rosidah *et al.*, 2009). In addition, Zahra *et al.* (2011) in their acute toxicity study has demonstrated no mortality with 5 g/kg dose of *G. procumbens* leaf extract in Sprague Dawley without showing any major clinical signs of toxicity (Zahra *et al.*, 2011).

Numerous *G. procumbens* products had been patented for various illness including gum protection, but the majority of them are for preparations of traditional Chinese medicine (Liao, 2015; Yang *et al.*, 2015).

## Chemistry

Considerable work has already been done to identify and isolate the chemical constituents from different extracts of *G. procumbens* (Kaewseejan *et al.*, 2008). Numerous studies have shown that various extract of *G. procumbens* leaves contains several active chemical constituents such as flavonoids, saponins, tannins, terpenoids and steroid glycosides. Previous studies had also reported that *G. procumbens* leaves extract contained rutin, kaempferol and two potential antioxidant components which are kaempferol-3-*O*-rutinoside and astragalin (**124**) (Figure 14.6) (Akowuah *et al.*, 2002). However, as far as we are aware, there is no bioactive compound isolated related to wound healing.

## *PANGIUM EDULE* REINW.

*Pangium edule* Reinw. or kepayang or keluak is a tall native Southeast Asian (including Indonesia and Papua New Guinea) tree species belonging to the family Flacourtiaceae or Acariaceae. It produces a large fruit (the 'football fruit') whereby its seed can be made edible by fermentation.

### Traditional Uses

The Iban of Sarawak apply the sap from the inner bark as anti-septic to treat wound (Chai, 2006).

### Biological Activity

Sap from *P. edule* induced fibroblast proliferation with a dose response at concentrations of 0.05 up to 5% v/v. and also inhibited MMP-9 (matrix mettaloproteinase 9) enzyme and SMAD signaling assays which contributed to a potential dual action effect (Prescott *et al.*, 2017). A characteristic of chronic wound environment is fibroblasts that exhibit premature senescence phenotype and plants that stimulate fibroblast growth which is essential for collagen deposition may assist in stimulating wound healing (Prescott *et al.*, 2017). At the same time, excessive protease activity from enzymes such as MMP-9 contribute towards stalled wound healing through the degradation of extracellular matrix and growth factors (Prescott *et al.*, 2017).

### Chemistry

No bioactive compounds have been isolated from *P. edule* so far that is responsible for wound healing.

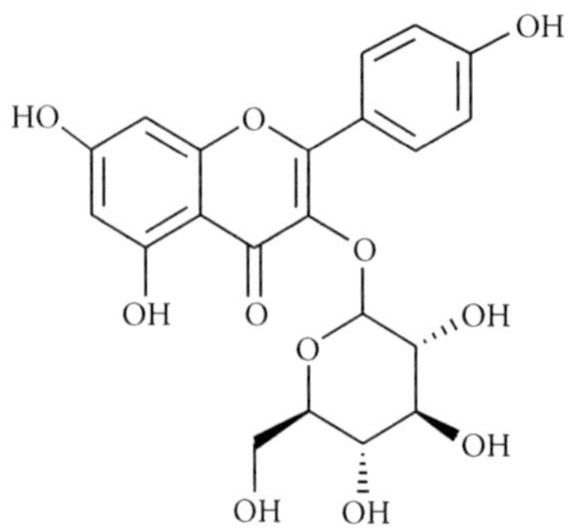

HO O OH O OH O O OH OH OH OH

**124**

**FIGURE 14.6** Astragalin (**124**), a flavonoid glycoside with anti-inflammatory properties.

## *Paspalum conjugatum* PJ.Berhius

*Paspalum conjugatum* PJ.Berhius (Figure 14.7) is a perennial tropical to subtropical grass and is originally from South America but has since been naturalized in many regions including Southeast Asia.

### Traditional Uses

The Iban of Borneo use leaf decoctions in the treatment of wounds and sores. The Kenyah of Sarawak use the juice from the leaves of *P. conjugatum* to stop bleeding by applying on the wound (Chai, 2006). Finely pounded leaves are tapped and rubbed on the wound caused by sharp object in Indonesian Borneo (Karmilasanti and Fernandes, 2017).

### Biological Activity and Chemistry

The presence of a hemostatic glucoside, paspaloside (3,3',4',5,7-Pentahydroxy-ó-rhamnosylñavone), which reduced the time for blood clotting by 50%, has been reported for this species.

## *Aleurites moluccanus* (L.) Willd.

*Aleurites moluccanus* (L.) Willd. (synonym: *Mallotus mollucanus*) (Figure 14.8) also called candlenut belongs to the Euphorbiaceae family. Its native range is impossible to establish precisely because of early spread by humans and the tree is now distributed throughout the new World and the Old World tropics.

**FIGURE 14.7** *Paspalum conjugatum* (Photo credit: Shuttlestock).

**FIGURE 14.8** *Aleurites moluccanus* (Photo credit: Shuttlestock).

### Traditional Use

Leaves are pounded and applied to wound by the Dayak Tunjung tribe in East Kalimantan (Setyowati, 2010)

### Biological Activity and Chemistry

By means of using carageenan induced rat paw oedema assay, *A. moluccanus* methanolic extract was shown to decrease the paw oedema at a dose of 300 mg/kg. The anti-inflammatory activity involved two phases in which the first phase was the result of the release of histamine and serotonin followed by the second phase whereby prostaglandin was released which resulted in oedema formation (Niazi *et al.*, 2010).

By incorporating a *A. moluccanus* ethanolic bark extract in simple ointment base at a concentration of 2% (w/w) and 4% (w/w) and topically applied to three types of model in rats viz. excision, incision and burn wound model, wound healing activity showed a similarity to those of a standard drug (nitrofurazone) in wound contracting ability, wound closure time and tensile strength (Prasad *et al.*, 2011). The wound applied with ointment containing 4% w/w alcoholic extract demonstrated significant ($p < 0.001$) wound contracting ability and period of epithelization while tensile strength was found to be significant with both the ointment formulations 2% w/w and 4% w/w. These were further supported and confirmed by the outcomes of histopathological examination of the excision and burn wound model (Prasad *et al.*, 2011).

## *Justicia gendarussa* Burm. f.

*Justicia gendarussa* Burm. f. (Acanthaceae) (Figure 14.9) is an evergreen shrub growing to 1 m in height. It has diverse biological activities and can grow in the open or semi-shade, preferably in moist soil.

### Traditional Uses

A poultice of the fresh leaves is traditionally used for the treatment of minor cuts, fractures, wounds and sprains by the tribes of Iban, Kayan and Kenyah in Sarawak.

### Biological Activity

The ethanol and water extracts of *J. gendarussa* leaves may have the potential to promote wound healing and antibacterial activities. Water and ethanol extracts of *J. gendarussa* leaves were evaluated for their wound healing activity in the form of an ointment (10% w/w) using the excision wound model. The ethanol extract in the form of ointment (10% w/w) exhibited a significant ($p < 0.01$) wound healing activity on the 14th day, and this is followed by water extract ($p < 0.05$) (Kumari *et al.*, 2016). Antibacterial activities against *Bacillus subtilis*, *Proteus mirabilis*, *S. aureus*, *Escherichia coli* and *Salmonella typhi* showed that among the two extracts,

**FIGURE 14.9** *Justicia gendarussa*. (Photo credit: Shuttlestock).

the ethanol extract exhibited good antibacterial activity with its MICs and MBCs at 1–2.5 mg/ml and 2–3 mg/ml, respectively. The MICs and MBCs of the formulated ointment were determined as 0.15–0.75 mg/ml and 1–1.5 mg/ml, respectively and comparable with standard antibiotic, bacitracin ointment (1 mg/g) (Kumari *et al.*, 2016).

Juheini and Elya (2010) determined the $LD_{50}$ value by the number of deaths in the group in 24 hours after giving a single dose of test substance. The result demonstrated that the highest dose was practically non-toxic, with an $LD_{50}$ value of 31.99 g/kg body weight (male rat groups) and 27.85 g/kg body weight (female rat groups). Another method was the measurement of aminotransferase activity done by using the colorimetric method. The outcome of ANOVA analysis for liver function showed that giving 4 g/kg body weight–16 g/kg body weight did not differ significantly between treated groups and control group (Juheini and Elya, 2010).

### Chemistry

The results confirmed the presence of carbohydrates, glycosides, alkaloids, flavonoid, phenols, tannins and saponin and the plant extracts show higher yield of total phenolic content and also flavonoid content (Muhameead *et al.*, 2019). Compounds such as flavonoids (Shamili and Santhi, 2019) and alkaloids (da Souza *et al.*, 2016) and aromatic amines (Kumar *et al.*, 1982) have been isolated from the leaves of *J. gendarussa*. However, no isolation of compounds with wound healing activity has ever been done.

In general, the medicinal plants used for wound healing can be considered as not very well studied (Table 14.1). An analysis using the available literature of the different experiments conducted on the above 10 species indicated that some species were better studied than others. This could be due to various reasons, such as whether the species is popular, seasonal, easily available (weedy and secondary species or ease of cultivation), widespread in usage and so forth. *Aleurites moluccanus and Nephelium lappaceum* are more restricted in use while *Melastoma malabathricum, Gynura procumbens* and *Justicia gendarussa* are more widely used and cultivated or found in secondary forest. But a more reliable indicator is probably the toxicity test, and only 3 out of 10 species or 30% have ever been studied for toxicity.

**TABLE 14.1**
**An Analysis of the Various Studies Done Related to Wound Healing**

| Species | Hemostasis | Proliferation | | | Remodeling | Toxicity |
|---|---|---|---|---|---|---|
| | | Wound filling | Wound contraction | Epithelization | | |
| *Nephelium lappaceum* L. | | | √ | | | |
| *Morinda citrifolia* L. | | √ | √ | √ | | √ |
| *Melastoma malabathricum* L. | | √ | √ | √ | √ | |
| *Ageratum conyzoides* (L.) L. | √ | √ | √ | √ | | √ |
| *Musa × paradisiaca* L. | | | √ | √ | | |
| *Gynura procumbens* (Lour.) Merr. | | √ | √ | | | |
| *Pangium edule* Reinw. | | √ | | | | |
| *Paspalum conjugatum* PJ.Berhius | √ | | | | | |
| *Aleurites moluccanus* (L.) Willd. | | | √ | √ | | |
| *Justicia gendarussa* Burm. *f.* | | √ | | | | √ |

# Conclusion

*Stephen P. Teo*

The well-known biodiversity of Borneo is also reflected in the diversity of medicinal plants used in Borneo that come from diverse plant families and taxa. These encompass both the dicotyledonous and monocotyledonous species, though the majority is dicotyledons.

All plant parts have been used for medicinal purposes including leaves, whole plants, stems, bark, roots, rhizomes and latexes, amongst others. Some such as member of the Guttiferae family (e.g. *Garcinia* and *Calophyllum*), Sapotaceae (e.g. *Palaquium*), Euphorbiaceae (e.g. *Macaranga*) as well as Apocynaceae (*Alstonia*) contain latexes or saps. The methods used in preparing the medicinal plants vary but the more commonly used methods include using tincture (aqueous extract) such as tea (e.g. gaharu tea) or the application of latexes or pounded plant materials such as leaves for wound.

The chemistry of the medicinal plants is highly diverse too and encompasses alkaloids, terpenoids, phenolics (e.g. flavonoids), polyketides and so forth. In terms of phytochemistry, certain species are better studied or investigated than the majority of the medicinal species. There is a dearth in information such as toxicities or cytotoxicities as well as the mechanisms of action of the bioactive compounds which are still poorly understood or completely lacking. Even the bioactive compounds of quite a sizeable medicinal species have not been isolated let alone the information on the syntheses of the bioactive compounds and studies using animal models.

More intensive research ought to be in place to further strengthen the uses of such medicinal plants and their contribution to the economy. Fundamental research areas such as phytochemistry, cytotoxicities and experimental animal studies and even clinical trials should be prioritized in order for the medicinal and herbal industry to move forward.

Documentation of medicinal plants and their traditional uses is important at this stage in Borneo before the knowledge is lost irretrievably as such information is mostly confined to elderly folks. Another concern is the threat from deforestation and land use change which can cause plant biodiversity, especially those rare and endemic species, to be lost forever.

# Bibliography


Adinortey, M. 2018. Wound healing potential of *Musa paradisiaca* L. (Musaceae) stem juice extract formulated into an ointment. *Research Journal of Pharmacology and Pharmacodynamics* 2011: 294–296.

Adinugraha, H. A. 2011. Pulai (*Alstonia scholaris* R. Br.). Info Tanaman Kehutanan. Yogyakarta: Balai Besar Penelitian Bioteknologi dan Pemuliaan Tanaman Hutan. (In Indonesian).

Ahmad, F. B. and Holdsworth, D. K. 1994. Medicinal plants of Sabah, Malaysia, part II. The Muruts. *International Journal of Pharmacognosy* 32(4): 378–383.

Ahmad, F. B. and Holdsworth, D. K. 1995. Traditional medicinal plants of Sabah, Malaysia part III. The Rungus people of Kudat. *International Journal of Pharmacognosy* 33(3): 262–264.

Ahmad, F. B. and Holdsworth, D. K. 2003. Medicinal plants of Sabah, East Malaysia–Part I. *Pharmaceutical Biology* 41(5): 340–346.

Ahmad R., Shaari K., Lajis, N. H., Hamzah, A. S., Ismail, N. H., *et al.* 2005. Anthraquinones from *Hedyotis capitellata*. *Phytochemistry* 66(10): 1141–1147.

Ahmad, W., Ibrahim J. and Syed Bukhari, N. A. 2016. *Tinospora crispa* (L.) Hook. *f.* & Thomson: A review of its ethnobotanical, phytochemical, and pharmacological aspects. *Frontiers in Pharmacology* 7: 59. doi: 10.3389/fphar.2016.00059. https://www.frontiersin.org/articles/10.3389/fphar.2016.00059/full (accessed on 22 July 2020).

Akowuah, G. A., Sadikum, A. and Mariam, A. 2002. Flavonoid identification and hypoglycaemic studies of the butanol fraction from *Gynura procumbens*. *Pharmaceutical Biology* 40: 405–410.

Al-Snafi, A. E. 2018. Chemical constituents, pharmacological effects and therapeutic importance of *Hibiscus rosa-sinensis* – A review. *Journal of Pharmacy* 8(7): 101–119.

Almeida, E. S., de-Oliveira, D. and Hotza, D. 2019. Properties and applications of *Morinda citrifolia* (Noni): A review. *Comprehensive Reviews in Food Science and Food Safety* 18(4): 883–909. doi.org/10.1111/1541-4337.12456.

Alves, A. T. M., Kloos, H., and Zani, C. L. 2003. Eleutherinone, a novel fungi toxic naphthoquinone from *Eleutherine bulbosa* (Iridaceae). *Memórias do Instituto Oswaldo Cruz* 98: 709–712.

Alwash, M. S. A., Ibrahim, N., Wan Ahmad, W. Y. and Din, L. B. 2014. Antibacterial, antioxidant and cytotoxicity properties of traditionally used *Melastoma malabathricum* Linn leaves. *Advance Journal of Food Science and Technology* 6(1): 6–12.

Amutha, K. and Selvakumari, U. 2016. Wound healing activity of methanolic stem extract of *Musa paradisiaca* Linn. (Banana) in Wistar albino rats. *International Wound Journal* 13(5): 763–767. doi: 10.1111/iwj.12371.

Ang, H. H., Chan, K. L. and Mak, J. W. 1995a. *In vitro* antimalarial activity of quassinoids from *Eurycoma longifolia* against Malaysian chloroquine-resistant *Plasmodium falciparum* isolates. *Planta Medica* 61(2): 177–178. doi.org/10.1038/nmicrobiol.2017.120.

Ang, H. H., Chan, K. L. and Mak, J. W. 1995b. Effect of 7-day daily replacement of culture medium containing *Eurycoma longifolia* Jack constituents on the Malaysian *Plasmodium falciparum* isolates. *Journal of Ethnopharmacology* 49(3): 171–175.

Anon. 2015. Global burden of diseases. 2015 data. *The Lancet* https://www.thelancet.com/gbd/2015 (accessed on 27 July 2020).

Anon. 2017. Stop neglecting fungi. *Nature Microbiology* 2(8): 17120 doi:10.1038/nmicrobiol.2017.120. ISSN 2058-5276 (accessed on 22 July 2020).

Anon. 2020. The Plant List. Version 1.1. http://www.theplantlist.org/ (accessed 12 February 2020).

Armania, N., Yazan, L. S., Musa, S. N., Ismail, I. S., Foo, J. B., *et al.* 2013a. *Dillenia suffruticosa* exhibited antioxidant and cytotoxic activity through induction of apoptosis and G2/M cell cycle arrest. *Journal of Ethnopharmacology* 146(2): 525–535.

Armania, N., Yazan, L. S., Ismail, I. S., Foo, J. B., Tor, Y. S., *et al.* 2013b. *Dillenia suffruticosa* extract inhibits proliferation of human breast cancer cell lines (MCF-7 and MDA-MB-231) via induction of G2/M arrest and apoptosis. *Molecules* 18(11): 13320–13339.

Arulprakash, K., Murugan, R., Ponrasu, T., Iyappan, K., Gayathri, V. S., *et al.* 2012. Efficacy of *Ageratum conyzoides* on tissue repair and collagen formation in rats. *Clinical and Experimental Dermatology* 37(4): 418–424. doi: 10.1111/j.1365-2230.2011.04285.x.

Athanasiou, K. A., Darling, E. M., Hu, J. C., DuRaine, G. D., and Reddi, A. H. (2013). Articular Cartilage. Boca Rato: CRC Press.

Azmi, L. 2011. Pharmacological and biological overview on *Mimosa pudica* Linn. *International Journal of Pharmacy and Life Sciences* 2(11): 1226–1234.

Babu, R. S. C., Rao, B., Babu, H. and Subbaraju, G. V. 2013. Isolation and bioactivity of diacetyltetritol from *Merremia emarginata* (Burm.*f*). *Natural Products Indian Journal* 9(5): 201–208.

Bach, Q. N., Hongthong, S., Quach, L. T., Pham, L., Pham, V., *et al.* 2018. Antimicrobial activity of rhodomyrtone isolated from *Rhodomyrtus tomentosa* (Aiton) Hassk. *Natural Product Research* 2: 1–6. doi: 10.1080/14786419.2018.1540479.

Badami, R. C. and Paul, K. B. 1981. Structure and occurrence of unusual fatty acids in minor seed oils. *Progress in Lipid Research* 19: 119–153.

Bagri, P., Ali, M., Aeri, V. and Bhowmi, M. 2016. Isolation and antidiabetic activity of new lanostenoids from the leaves of *Psidium guajava* L. *International Journal of Pharmacy and Pharmaceutical Sciences* 8(9): 975-149.

Balamurugan, K., Nishanthini, A. and Mohan, V. R. 2013. Antiulcer activity of *Melastoma malabathricum* L. leaf extracts (Melastomataceae). *International Journal of Advanced Research* 1(5): 49–52 49.

Baling, J., Noweg, G. T., Sayok, A. K., Wadell, I. and Ripe, J. E. 2017. Medicinal plants usage of Jagoi Bidayuh Community, Bau District, Sarawak, Malaysia. *Journal of Borneo Kalimantan* 3(1): 67–87.

Bamidele, O., Akinnuga, A. M., Anyakudo, M. M. C., Ojo, O. A., Ojo, G. B., *et al.* 2010. Haemostatic effect of methanolic leaf extract of *Ageratum conyzoides* in albino rats. *Journal of Medicinal Plants Research* 4(20): 2075–2079.

Bantol, J. G., Berdin, N. M. R., Pernascoza, Z. P., and Abrea, J. A. A. 2017. *In vitro* anti-inflammatory assays on hexane extract of sambong (*Blumea balsamifera*) leaves. 7th Cebu International Conference on Civil, Agricultural, Biological and Environmental Sciences (CABES-17) 21–22 September 2017, Cebu, Philippines.

Barnard, D. L., Huffman, J. H., Wood, S. G., and Hughes, B. G. 1992. Evaluation of the antiviral activity of anthraquinones, anthrones and anthraquinone derivatives against human cytomegalovirus. *Antiviral Research* 17(1): 63–77.

Begum, Z., Younos, I. and Khan, H. 2018. Analgesic and anti-inflammatory activities of the ethanol extract of *Hibiscus rosa-sinensis* Linn. (roots). *Pakistan Journal of Pharmaceutical Sciences* 32(5): 1927–1933.

Bertani, S. G., Bourdy, I., Landau, J. C., Robinson, P. H., Esterre, and Deharo, E. 2005. Evaluation of French Guiana traditional antimalarial remedies. *Journal of Ethnopharmacology* 98(1–2): 45–54.

Bhattarai, K., Pokharel, B., Maharjan, S. and Adhikari, S. 2018. Chemical constituents and biological activities of ginger rhizomes from three different regions of Nepal. *Journal of Nutritional Dietetics and Probiotics* 1(1): 1–12.

Bi, K., Touré, D., Kablan, L., Bedi, G., and Tea, I. 2018. Medicinal plant of Ivory Coast: chemical and biological study. *Records of Natural Products* 12(2): 160–168. doi: 10.25135/rnp.22.17.06.040.

Biedenkopf, N., Lange-Grünweller K., Schulte, F. W., Weißer. A., Müller, C., *et al.* 2016. The natural compound silvestrol is a potent inhibitor of Ebola virus replication. *Antiviral Research* 137: 76–81. doi: 10.1016/j.antiviral.2016.11.011.

Birari, R. B., Jalapure, S. S., Chnagrani, S. R., Shid, S. J., Mtote, M. V., *et al.* 2009. Anti-inflammatory, analgesic and antipyrectic effect of *Hibiscus rasa-sinensis* Linn. flower. *International Journal of Pharmacy and Pharmaceutical Sciences* 3: 737–747.

Buckheit, R. W. Jr., Russell, J. D., Xu, Z. Q., and Flavin, M. 2001. Anti-HIV-1 activity of calanolides used in combination with other mechanistically diverse inhibitors of HIV-1 replication. *Antiviral Chemistry and Chemotherapy* 11(5): 321–277.

Burdick, E. M. 1971. Carpaine: An alkaloid of *Carica Papaya*: Its chemistry and pharmacology. *Economic Botany* 25(4): 363–365.

Cabot, M. C. and Goucher, C. R. 1981. Chaulmoogric acid: Assimilation into the complex lipids of mycobacteria. *Lipids* 16(2): 146–148.

Cachet, X., Langrand, J., Riffault-Valois, L., Bouzidi, C., Colas, *et al.* 2018. Clerodane furanoditerpenoids as the probable cause of toxic hepatitis induced by *Tinospora crispa*. *Scientific Reports* 8(1): 13520–13526.

Caetano, C. P., de Sá, C. B., Faleiros, B. A. P., Gomes, M. F. C. F. and Pereira, E. R. S. 2017. Neurotoxicity following the ingestion of bilimbi fruit (*Averrhoa bilimbi*) in an end-stage renal disease patient on hemodialysis. *Case Reports in Nephrology and Dialysis* 7(1): 6–12.

Centre for Disease Control. 2018. Malaria's Impact Worldwide. Published on the Internet; https://www.cdc.gov/malaria/malaria_worldwide/impact.html (accessed 3 November 2019).

Cencic, R., Carrier, M., Galicia-Vázquez, G., Bordeleau, M. E., Sukarieh, R., *et al.* 2009. Antitumor activity and mechanism of action of the cyclopenta [b] benzofuran, silvestrol. *PloS one*, 4(4), p. e5223. https://journals.plos.org/plosone/article/comments?id=10.1371/journal.pone.0005223 (accessed on 22 July 2020).

Chai, P. K. K. 2000. A list of medicinal plants from the Kedayan communities. In: A Checklist of flora, fauna, food and medicinal plants at Lanjak Entimau Wildlife Sanctuary, ed. Chai, 157–170. Kuching: Forest Department Sarawak, Malaysia and International Tropical Timber Organization (ITTO), Japan.

Chai, P. P. K. 2006. *Medicinal plants of Sarawak*. Kuching: Forest Department Sarawak.

Chan, K. L., O'Neill, M. J., Phillipson, J. D. and Warhurst, D. C. (1986). Plants as sources of antimalarial drugs. Part 31 *Eurycoma longifolia*. *Planta Medica* 52(2): 105107.

Chang, F. R. and Wu, Y. C. 2001. Novel cytotoxic annonaceous acetogenins from *Annona muricata*. *Journal of Natural Products* 64(7): 925–931.

Chao, W.-W. and Lin, B.-F. 2010. Isolation and identification of bioactive compounds in *Andrographis paniculata* (Chuanxinlian). *Chinese Medicine* 5. https://cmjournal.biomedcentral.com/articles/10.1186/1749-8546-5-17 (accessed on 22 July 2020).

Charles, R., Garg, N. and Kumar, S. 2000. New gingerdione from the rhizome of *Zingiber officinale*. *Fitoterapia* 71(6): 716–718.

Chen, C. K., Lin, F. H., Tseng, L. H., Jiang, C. L. and Lee, S. S. 2011. Comprehensive study of alkaloids from *Crinum asiaticum* var. *sinicum* assisted by HPLC-DAD-SPE-NMR. *Journal of Natural Products* 74(3):411–419. doi: 10.1021/np100819n. Epub 2011 Feb 11.

Chena, D., Qiao, J., Sun, Z., Liu, Y., and Sun, Z. 2019. New naphtoquinones derivatives from the edible bulbs of *Eleutherine americana* and their protective effect on the injury of human umbilical vein endothelial cells. *Fitoterapi* 132: 46–52.

Chenera, B., West, M. L., Finkelstein, J. A. and Dreyer, G. B. 1993. Total synthesis of calanolide A, a non-nucleoside inhibitor of HIV-1 reverse transcriptase. *Journal of Organic Chemistry* 58: 5605–5606.

Choi, J., Tai, H., Cuong, M., Kim, H., and Jang, D. 2012. Antioxidative and antiinflammatory effect of quercetin and its glycosides isolated from mampat (*Cratoxylum formosum*). *Food Science and Biotechnology* 21(2): 587–595.

Chouni, A. and Paul, S. 2018. A Review on Phytochemical and Pharmacological Potential of Alpinia galanga. *Pharmacognosy Journal* 10(1): 9–15.

Christensen, H. 2002. Ethnobotany of the Iban and the Kelabit. Aarhus: Forest Department Sarawak, Malaysia, Nature, Ecology and People Consult (NEPCon) and University of Aarhus.

Christina Injan, A. M. 2017. Antibacterial and anti-biofilm activities of *Dicranopteris linearis* leaf extracts. PhD diss., Monash University Malaysia, Malaysia.

Coleman, S. L. 1997. Only plant-based anti-HIV agent goes into clinical trial. Arnold Arboretum. https://www.newswise.com/articles/only-plant-based-anti-Hiv-agent-goes into-clinical-trial (accessed on 20 March 2019).

Connell, D. W. and McLachlan, R. 1972. Natural pungent compounds: IV. Examination of the gingerols, shogaols, paradols and related compounds by thin-layer and gas chromatography. *Journal of Chromatography* 67(1): 29–35.

Creagh, T., Ruckle, J. L., Tolbert, D. T., Giltner, J., Eiznhamer, D. A., *et al.* 2001. Safety and pharmaacokinetics of single doses of (+)-calanolide a, a novel, naturally occurring nonnucleoside reverse transcriptase inhibitor, in healthy, human immunodeficiency virus-negative human subjects. *Antimicrobial Agent and Chemotherapy* 45(5): 1379–1386.

Currens, M. J., Gulakowski, R. J., Mariner, J. M., Moran, R. A., Buckheit, R. W. Jr., *et al.* 1996a. Antiviral activity and mechanism of action of calanolide A against the human immunodeficiency virus type-1. *Journal of Pharmacology Experimental Therapy* 279(2): 645–651.

Currens, M. J., Mariner, J. M., McMahon, J. B., and Boyd, M. R. 1996b. Kinetic analysis of inhibition of human immunodeficiency virus type-1 reverse transcriptase by calanolide. *American Journal of Pharmacology and Experimental Therapeutics* 279: 652–661.

Daud, D., Fazuliana, N. U. R., Arsad, M., Ismail, A. and Tawan, A. 2016. Anti-pyrectic action of *Caulerpa lentillifera*, *Hibiscus rosa-sinensis* and *Piper sarmentosum* aqueous extract in mice. *Asian Journal of Pharmaceutical and Clinical Research* 9(1): 9–11.

da Souza, L. G., Almeida, M. C. S., Lemos. T. L. G. and Riceli, P. 2016 Brazoides A-D, new alkaloids from *Justicia gendarussa* Burm. *f.* species. *Journal of the Brazilian Chemical Society* 28(7): 1–5. doi: 10.21577/0103-5053.20160291.

Dayie, N., Newman, N., Ayitey-Smith and Tayman, E. F. 2007. Screening for antimicrobial activity of *Ageratum conyzoides* L.: A pharmaco-microbiological approach. *The Internet Journal of Pharmacology*. 5(2). http://ispub.com/IJPHARM/5/2/9551 (accessed on 22 July 2020).

Devi, P. S., Rukmini, K., Indrani, V. and Devamma, M. N. 2015. Antimicrobial studies and identification of cellular components of *Dicranopteris linearis* from Tirumala Hills. *International Journal of Pharmceutical Research and Review* 4(8): 13–17. ISSN: 2278-6074.

Diba, F., Yusro, F., Matiani, Y. and Ohtani, K. 2013. Inventory and biodiversity of medicinal plants from tropical rain forest based on traditional knowledge by ethnic Dayaknese communities in West Kalimantan, Indonesia. *Kuroshio Science* 7(1): 75–80.

dos Anjos, P. J. C., Pereira, P. R., Moreira, I. J. A., Serafini, M. R., Araujo, A. A. S., *et al.* 2013. Antihypertensive effect of *Bauhinia forficata* aqueous extract in rats. *Journal of Pharmacology and Toxicology* 8(3): 82–89.

dos Santos, O. A. C., Souza, D. S., Mesquita, T. R. R., de Menezes-Filho, R. J. E., Caldas, A. P. D., *et al.* 2018. *Averrhoa bilimbi* L. leaf aqueous extract modulates both cardiac contractility and frequency in the guinea-pig atrium by the activation of muscarinic receptors. *Letters in Drug Design and Discovery* 15(11): 1163–1169. doi:10.2174/1570180815666180125150457.

Duc, L. V., Thanh, T. B., Giang, N. P. and Tien, V. N. 2017. Anti-inflammatory and anti-cancer activities of *Hedyotis capitellata* growing inVietnam. *World Journal of Medical Sciences* 14(2): 22–28.

Duong, N. T. T., Chinh, H. T., Din,T. S., Phong, T. L. H. and Phuong, P. N. 2013. Contribution to the study on chemical lconstituents from the leaves of *Cassia alata* L., (Caesalpiniaceae). *Science and Technology Development* 16(2): 26–31.

Echeverri, F., Torres, F., Quinones, W., Cardona, G., Archbold, R., *et al.* 1997. Danielone, a phytoalexin from papaya fruit. *Phytochemistry* 44(2): 255–256.

Elumalai, E., Muthiyan, R., Thirunavukkarasu, T. and Vinothkumar, P. 2011. Antibacterial activity of various leaf extracts of *Merremia emarginata*. *Asian Pacific Journal of Tropical Biomedicine* 1(5): 406–408.

Falah, F., Sayektiningsih T. and Noorcahyati 2013. Keragaman jenis dan pemanfaatan tumbuhan berkhasiat obat oleh masyarakat sekitar hutan lindung Gunung Beratus, Kalimantan Timur. *Jurnal Penelitian Hutan dan Konservasi Alam* 10(1): 1–18. (abstract in English).

Farahpour, M. R. 2019. Medicinal plants used in wound healing. *IntechOpen* doi 10.5772/interchopen80215. https://www.intechopen.com/books/wound-healing-current-perspectives/medicinal-plants-in-wound-healing (accessed 20 March 2020).

Farazimah, H. J. Y., Norhayati, A., Mohamad, A. M., Sivasothy, Y., Khalijah A., *et al.* 2018. Isolation of a novel compound from *Dillenia suffruticosa* (Griff) Mart. Joint Event on 4th World Congress on Medicinal Plants and Natural Products Research and 12th Global Ethnomedicine and Ethnopharmacology Conference, 8–9 August 2018, Osaka, Japan.

Fernand, V. E., Dinh, D. T., Washington, S., Fakayode, S. and Losso, J. N. 2008. Determination of pharmacologically active compounds in root extracts of *Cassia alata* L. by use of high performance liquid chromatography. *Talanta* 74(4): 896–902. doi:10.1016/j.talanta.2007.07.033.

Fouedjou, R. T., Teponno, R. B., Quassinti, L., Bramucci, M., Petrel, I. D., *et al.* 2014. Steroidal saponins from the leaves of *Cordyline fruticosa* (L.) A. Chev. and their cytotoxic and antimicrobial activity. *Phytochemistry Letters* 7: 62–68.

Galinis, D. L., Fuller, R. W., McKee, T. C., Cardellina, J. H., Gulakowski, R. J., *et al.* 1996. Structure activity modifications of the HIV-1 inhibitors (+) Calanolide A and (−)-Calanolide B. *Journal of Medicinal Chemistry* 39: 4507–4510.

Gandhi, M. and Vinayak, V. K. 1990. Preliminary evaluation of extracts of *Alstonia scholaris* bark for *in vivo* antimalarial activity in mice. *Journal of ethnopharmacology* 29(1): 51–57.

Gbenou, J. D., Ahounou, J. F., Akakpo, H. B., Laleye, A., Yayi, E., *et al.* 2013. Phytochemical composition of *Cymbopogon citrates* and *Eucalyptus citriodora* essential oils and their anti-inflammatory and analgesic properties in Wistar rats. *Molecular Biology Reports* 40(2): 1127–1134.

Genest, S. 2008. Comparative bioactive studies on two *Mimosa* species. *Boletin Lationoamericano y del Caribe de Plantas Medicinale y Aromaticas* 7(1): 102–340.

Goh, M. P. Y., Basri, A. M., Yasin, H., Taha, H. and Ahmad, N., 2017. Ethnobotanical review and pharmacological properties of selected medicinal plants in Brunei Darussalam: *Litsea elliptica, Dillenia suffruticosa, Dillenia excelsa, Aidia racemosa, Vitex pinnata and Senna alata. Asian Pacific Journal of Tropical Biomedicine* 7(2): 173–180.

Govindachari, T. R., Pai, B. R. and Narasimhan, N. S. 1954. Pseudocarpaine, a new alkaloid from *Carica papaya* L. *Journal of Chemical Society* 1954: 1847–1849.

Gritsanapan, W., Wuthi-udomlert, M., Tridej, M. and Archawakom 1998. Study of antifungal activity of anthraquinones from Cassia alata Linn. leaves. Congress on Science and Technology of Bangkok (Thailand). 19–21 October 1998, Bangkok, Thailand.

Grünweller, A. and Hartmann, R. K. 2017. Silvestrol: A potential future drug for acute Ebola and other viral infections. *Future Medicine* 11: 243–245.

Guo, S. and DiPietro, L. A. 2010. Factors affecting wound healing. *Journal of Dental Research* 89(3): 219–229. doi: 10.1177/0022034509359125.

GRAIN. 2017. Sarawak MediChem tests proprietary plant-based anti-HIV agent. https://www.grain.org/article/entries/2039-sarawak-medichem-tests-proprietary-plant-basedanti-hiv-agent (accessed on 22 July 2020).

Gupta, P. and Birdi, T. 2015. *Psidium guajava* leaf extract prevents intestinal colonization of *Citrobacter rodentium* in the mouse model. *Journal of Ayurveda and Integrative Medicine* 6: 50–52.

Hadi, S. 2009 Mataranine A and B: A new diastomeric indole alkaloid from *Alstonia scholaris* R.Br. of Lombok Island, Indonesia. *Journal of Chemistry* 9(3): 505–508.

Hansra, D. M., Silva, O., Mehta, A. and Ahn, E. 2014. Patient with metastatic breast cancer achieves stable disease for 5 years on graviola and xeloda after progressing on multiple lines of therapy. *Advances in Breast Cancer Research* 3: 84–87.

Haraguch, H., Kuwata, Y., Inada, K., Shingu, K., Miyahara, K., *et al.* 1996. Antifungal activity from *Alpinia galanga* and the competition for incorporation of unsaturated fatty acids in cell growth. *Planta Medica* 62(4): 308–313.

Hay, R. J., Johns, N. E., Williams, H. L. C., Bolliger, I. W., and Dellavalle, R. P. 2013. The global burden of skin disease in 2010: An analysis of the prevalence and impact of skin conditions. *The Journal of Investigative Dermatology* 134(6): 1527–1534. doi:10.1038/jid.2013.446. PMID 24166134.

Hemlata, S. B. K. 1993. Alatinone, an anthraquinone from *Cassia alata*. *Phytochemistry* 32(6): 1616–1617.

Hennebelle, T., Weniger, B., Joseph, H. and Sahpaz, S. 2009. *Senna alata*. *Fitoterapia* 80 (7): 385–393.

Hettige, S. 2015. Guidelines in using *Carica papaya* leaf extract for Dengue fever patients. *British Medical Journal*. https://www.bmj.com/content/351/bmj.h4661/rr-4 (accessed on 06 Januaryr 2020).

Hirai, N., Ishida, H. and Koshimizu, K. 1994. A phenalenone-type phytoalexin from *Musa acuminata*. *Phytochemistry* 37 (2): 383–385.

Hoffman, B. R., DelasAlas, H., Blanco, K., Wiederhold, N., and Lewis, R. E. 2004. Screening of antibacterial and antifungal activities of ten medicinal plants from Ghana. *Pharmaceutical Biology* 42(1): 13–17.

Hossain, M., Islam, M. M., Azad, M. A. K., Al Faruq, A., Tareq, S. M., *et al.* 2016. *In vivo* investigation of analgesic, antipyretic, anti-diarrheal and anxiolytic activity of Blumean densiflora DC. *European Journal of Medicinal Research* 3(11): 50–55.

Huong, T. T., Cuong, N. X., Tram, L. H., Quang, T. T., Duong, L. V., *et al.* 2012. A new prenylated aurone from *Artocarpus altilis*. *Journal of Asian Natural Products Research* 14(9): 923–928. doi:10.1080/10286020.2012.702758.

Ifesan, B. O. T., Joycharat, N. and Voravuthikunchai, S. P. 2009. The mode of anti-staphylococcal action of *Eleutherine americana*. *FEMS Immunology and Medical Microbiology* 57: 193–201.

Ifesan, B. O. T., Ibrahim, D. and Voravuthikunchai, S. P. 2010. Antimicrobial activity of crude ethanolic extract from *Eleutherine americana*. *Journal of Food, Agriculture and Environment* 8: 1233–1236.

Igboasoiyi, A. C., Eseyin, O. A., Ezenwa, N. K. and Oladimeji, H. O. 2007. Studies on the toxicity of *Ageratum conyzoides*. *Journal of Pharmacology and Toxicology* 2: 743–747.

Ihwan, I., Rifa'i, M. and Fitri, L. E. 2014. Antiplasmodial test of *Tinospora crispa* stem extract against *Plasmodium falciparum* 3D7 strain *in vitro*. *Jurnal Kedokteran Brawijaya* 28(2): 91–96.

Imam, M. Z. and Akter, S. 2011. *Musa paradisiaca* L. and *Musa sapientum* L.: A phytochemical and pharmacological review. *Journal of Applied Pharmaceutical Science* 1(5): 14–20.

Inada, A. C., Figueiredo, P. S., dos Santos-Eichler, R. A., de Freitos, K., *et al.* 2017. *Morinda citrifolia* Linn. (Noni) and its potential in obesity-related metabolic dysfunction. *Nutrients* 9(6): 1–29.

Insanu, M., Kusmardiyania, S., and Hartati, R. 2014. Recent studies on phytochemicals and pharmacological effects of *Eleutherine americana* Merr. *Procedia Chemistry* 13: 221–228.

Irfan, R. and Desy, S. 2018. Efficacy of mengkudu leaves extract (*Morinda citrifolia*) on wound healing rate post tooth extraction in white rats (*rattus norvegicus*). *Journal of Dentomaxillofacial Science* 3(1): 28–31. https://jdmfs.org/index.php/jdmfs/article/view/547 (accessed on 01 March 2020).

Ismail, A. F. H., Abd Samah, O. and Sule, A. 2011. A preliminary study on antimicrobial activity of *Imperata cylindrica*. *Borneo Journal of Resource Science and Technology* 1: 63–66.

Jacobsen, P. L. and Levy, L. 1973. Antimicrobial agents and chemotherapy, mechanism by which hydnocarpic acid inhibits mycobacterium multiplication. *American Society for Microbiology* 3(3): 373–379.

Jacobsen, P. L., Ng, H. and Levy, L. 1973. The susceptibility of mycobacteria to hydnocarpic acid. *American Review of Respiratory Disease* 107(6): 1022–1029.

Janssen, A. M. and Scheffer, J. J. 1985. Acetoxychavicol acetate, an antifungal component of *Alpinia galanga*. *Planta Medica* 1(6): 507–511.

Ji, S., Zhang, G., Hua, Y. and Jin, X. 2015. *Sanguis draconis* (*Daemonorops draco*). A case report of treating a chronic pressure ulcer with tunneling. *Holistic Nursing Practice* 29(1): 48–52.

Jiwajinda, S., Santisopasri, V., Murakami, A., Kawanaka, M. and Kawanaka, H. 2002. *In vitro* anti-tumor promoting and anti-parasitic activities of the quassinoids from *Eurycoma longifolia*, a medicinal plant in Southeast Asia. *Journal of Ethnopharmacology* 82(1): 55–58.

Joffry, S. M., Yob, N. J., Rofiee, M. S.,. Affandi, M. M., Suhaili, Z., *et al.* 2012. *Melastoma malabathricum* (L.) Smith ethnomedicinal uses, chemical constituents, and pharmacological properties. *Evidence-Based Complementary and Alternative Medicine* 2012: 258434–258449.

Johari, J., Kianmehr, A., Mustafa, R., Abubakar, S. and Zandi, K. 2012. Antiviral activity of baicalein and quercetin against the Japanese encephalitis virus. *International Journal of Molecular Sciences* 13(12): 16020–16045.

Jong, A. 2019. Rajah Brooke Memorial Hospital, Kuching. Personal communication.

Jong-Anurakkun, N., Bhandari, M. R. and Kawabata, J. 2007. α-Glucosidase inhibitors from Devil tree (*Alstonia scholaris*). *Food Chemistry* 103(4): 1319–1323.

Juheini, A. and Elya, B. 2010. Acute toxicity of *Justicia gendarussa* Burm. Leaves 14(2): 129–134.

Kaewseejan, N., Puangpronpitag, D. and Nakornriab, M. 2008. Evaluation of phytochemical composition and antibacterial property of *Gynura procumbens*. *Asian Journal of Plant Sciences* 2012: 1–5.

Kamiya, K., Hamabe, W., Tokuyama, S., Hirano, K., Satake, T., *et al.* 2010. Inhibitory effect of anthraquinones isolated from the Noni (*Morinda citrifolia*) root on animal A-, B-

and Y- families of DNA polymerases and human cancer cell proliferation. *Food Chemistry* 118(3): 725–730.

Kanokmedhakul, S., Kanokmedhakul, K., Kantikeaw, I. and Phonkerd, N. 2006. 2-Substituted furans from the roots of *Polyalthia evecta*. *Journal of Natural Products* 69(1): 68–72.

Kardono, L. B. S., Angerhofer, C. K., Tsauri, S., Admawinata, K., Pezzuto, J. M. and Kinghorn, A. D. Y. 1991. Cyotoxic and antimalarial constituents of the roots of *Eurycoma longifolia*. *Journal of Natural Products* 54(5): 1360–1361.

Karmilasanti dan Fernandes, A. 2013. Tumbuhan Obat Luka dari Tana Ulen. Seminar Peranan Hasil Litbang Hasil Hutan Bukan Kayu Dalam Mendukung Pembangunan Kehutanan. Kementrian Kehutanan, Badan Penelitian dan Pengembangan Kehutanan, Pusat Penelitian Dan Pengembangan Peningkatan Produktivitas Kehutanan, Indonesia. 345–356. https://www.researchgate.net/publication/301787441 (In Indonesian).

Kartikawati, S. M., Ervizal, A. M. Z., Hikmat, A., Kartodihardjo, H. and Fuadi, M. 2014. Habitat preferences, distribution pattern, and root weight estimation of pasakbumi (*Eurycoma longifolia* Jack.). *Jurnal Manajemen Hutan Tropika* 20(1): 43–50.

Keawpradub, N., Kirby, G. C., Steele, J. C. P. and Houghton, P. J. 1999. Antiplasmodial activity of extracts and alkaloids of three *Alstonia* species from Thailand. *Planta medica* 65(8): 690–694.

Khan, F. U. and Juliana, M. J., and Bakhtiar, M. T. 2013. *In-vitro* antifungal and antibacterial activities of rhizomes extracts of *Alpinia galanga* and *Alpinia conchigera* Griff. International Conference on Natural Products 2014, 18–19 March 2014, Palm Garden Hotel, IOI Resorts, Putrajaya, Malaysia.

Khoswanto, C. 2010. Mengkudu (*Morinda citrifolia* Linn.) gel affect on post-extraction fibroblast acceleration. *Dental Journal* 43(1): 31–34.

Khyade, M. S. and Vaikos, N. P. 2009a. Phytochemical and antibacterial properties of leaves of *Alstonia scholaris* R. Br. *African Journal of Biotechnology* 8: 6434–6436.

Khyade, M. S. and Vaikos, N. P. 2009b. Comparative phytochemical and antibacterial studies on the bark of *Alstonia scholaris* R.Br. and *Alstonia macrophylla* Wall. *Ex G.Don. Phamacognosy* 1(4): 246–249.

Kim, S., Hwang, B. Y., Su, B. N., Chai, H., Mi, Q., *et al.* 2007. Silvestrol, a potential anticancer rocaglate derivative from *Aglaia foveolata*, induces apoptosis in LNCaP cells through the mitochondrial/apoptosome pathway without activation of executioner caspase-3 or-7. *Anticancer Research* 27(4B): 2175–2183.

Kumar, A., Partha, C., Ghosh, P., Satyesh, D. and Pakrashi, C. 1982. Simple aromatic amines from *Justicia gendarussa*. $^{13}$C NMR spectra of the bases and their analogues. *Tetrahedron* 38(12): 1797–1802.

Kumari, A. V. A. G., Sunilson, A. J. J., Anandarajagopal, K. and Khan, A. 2016. Antibacterial and wound healing activities of *Justicia gendarussa* leaf extracts. 2nd Global Summit on Herbals and Natural Remedies. 17–19 October 2016, Kuala Lumpur, Malaysia.

Kuo, P.-C., Shi, L.-S., Damu, A.-G., Su, C.-R., Huang, C.-H., *et al.* 2003. Cytotoxic and antimalarial β-carboline alkaloids from the roots of *Eurycoma longifolia*. *Journal of Natural Products* 66(10): 1324–1327. https://doi.org/10.1021/np030277n.

Kuo, P.-C., Amooru, G. D., Lee, K.-H. and Wu, T.-S. 2004. Cytotoxic and antimalarial constituents from the roots of *Eurycoma longifolia*. *Bioorganic and Medicinal Chemistry* 12(3): 537–544.

Kuo, S.-C., Sun, H.-D., Morris-Natschke, S. L., Lee, K.-H. and Wu, T.-S. 2014. Cytotoxic cardiac glycosides and coumarins from *Antiaris toxicaria*. *Bioorganic Medicinal Chemistry* 22(6): 1889–1898. doi: 10.1016/j.bmc.2014.01.052.

Kustiawan, W. 2007. Medicinal plants of Kalimantan Forest: A review. *Natural Life* 2(1): 24–34.

Kuswantoro, F. 2017 Traditional antimalaria plants species of Balikpapan Botanic Garden, East Kalimantan. *KnE Life Sciences* 3(4): 78–85.

Lam, S.-H., Ruan, C.-T., Hsieh, P.-H., Su, M.-J. and Lee, S. S. 2012. Hypoglycemic diterpenoids from *Tinospora crispa*. *Journal of Natural Products* 75(2): 153–159.

Lambert, J. B., Levy, A. J., Niziolek, L. C., Feinman, G. M., Gayford, P. J., Santiago-Blay, J. A. and Wu, Y. 2016. The resinous cargo of the Java Sea wreck. *Archaeometry* 59(5): 1–16.

Lavanya, G., Voravuthikunchai, P. S. and Towatana, N. H. 2012. Acetone extract from *Rhodomyrtus tomentosa*: A potent natural antioxidant. *Evidence-Based Complementary and Alternative Medicine* 2012: 535479–535487.

Lay, D., Sugirtama, I. W. and Arjana, I. G. K. N. 2019. The effects of noni (*Morinda citrifolia*) ethanol extract cream on collagen deposition in incisional wound healing of male Wistar rats. *International Journal of Science and Technology* 8(5): 419–425.

Leaman, D. J., Arnason, J. J. T., Yusuf, R., Sangat-Roemantyo, H., Soedjito, H., *et al.* 1995. Malaria remedies of the Kenyah of the Apo Kayan, East Kalimantan, Indonesian Borneo: A quantitative assessment of local consensus as an indicator of biological efficacy. *Journal of Ethnopharmacology* 49(1): 1–16.

Leaman, D. J. 1996. The medicinal ethnobotany of the Kenyah of East Kalimantan (Indonesian Borneo). PhD diss., University of Ottawa, Canada.

Lee, S. H., Lee, S. Y., Son, D. J., Lee, H., Yoo, H. S., *et al.* 2005. Inhibitory effect of 2-hydroxycinnamaldehyde on nitric acid production through inhibition of NF-kB activation in RAW 253.7 cells. *Chemical Pharmacology* 69: 791–199.

Lestari, I., Melania, A. and Prasetyo, B. (2018). Potency water stew of *Averrhoa bilimbi* L for antihypertensive. *International Journal of Nursing and Midwifery Science* 2(1): 59–61.

Levy, L. 1975. The activity of chaulmoogra acids against *Mycobacterium leprae*. *American Review of Respiratory Disease* 111(5): 703–705.

Li, C., Lee, D., Graf, T. N., Phifer, S. S., Nakanishi, Y., *et al.* 2005. A hexacyclicenttrachylobane diterpenoid possessing an oxetane ring from *Mitrephora glabra*. *Organic Letters* 7(25): 5709–5712. doi: 10.1021/ol052498l.

Li C., Lee D., Graf, T. N., Phifer, S. S., Nakanishi Y., *et al.* 2009. Bioactive constituents of the stem bark of *Mitrephora glabra*. *Journal of Natural Products* 72(11): 1949–1953. doi: 10.1021/np900572g. PMID: 19874044; PMCID: PMC2862477.

Liao, G. 2015. Traditional Chinese medicinal composition for treating uterine cancer and its preparation method and application. CN. Patent No. 105055807. State Intellectual Property Office of the P.R.C., Beijing.

Limsuwan, S., Subhadhirasakul, S. and Voravuthikunchai, S. P. 2009. Medicinal plants with significant activity against important pathogenic bacteria. *Pharmaceutical Biology* 47: 683–689.

Lin, L.-Z., Hu, S. F., Chai, H. B., Pengsupar, T., Pezzuto, J. M., *et al.* 1995. Lycorine alkaloids from *Hymenocallis littoralis*. *Phytochemistry* 40(4): 1295–1298.

Little, Jr., Elbert, L. and Skolmen, R. G. 1989. Hau, sea hibiscus. Common forest trees of Hawaii (Native and Introduced). United State Forest Service.

Liu, Y., Liu, J. and Zhang, Y. 2019. Research progress on chemical constituents of *Zingiber officinale* Roscoe. *Biomed Research International* 2019: 1–21 https://doi.org/10.1155/2019/5370823.

Lokman, F. E., Gu, H. F., Wan Mohamud, W. N., Yusoff, M. M., Chia, K. L., *et al.* 2013. Antidiabetic effect of oral borapetol B compound, isolated from the plant *Tinospora crispa*, by stimulating insulin release. *Evidence-Based Complementary and Alternative Medicine* 2013: 602–622.

Luis, J. G., Echeverri, F., Quinones, W., Brito, I., Lopez, M., *et al.* 1993. Irenolone and emenolone: Two new types of phytoalexin from *Musa paradisiaca*. *Journal of Organic Chemistry* 58(16): 4306–4308.

Mai, N. T. T., Hai, N. X., Phu, D. H., Trong, P. N. H. and Nhan, N. T. 2012. Three new geranyl aurones from the leaves of *Artocarpus altilis*. *Phytochemistry Letter* 5(3): 647–757.

Manh, H. D., Ngoc, T. M. and Ahn, J. S. 2017. Inhibitory activity of diterpenoids isolated from *Tetracera scandens*. *Journal of Medicinal Materials* 22(5): 283–287.

Manivannan, J., Shanthakumar, J., Silambarasan, T., Balamurugan, E. and Raja, B. 2015. Diosgenin, a steroidal saponin, prevents hypertension, cardiac remodeling and oxidative stress in adenine induced chronic renal failure rats. *RSC Advances* 25: 19337–19344.

Manoj, G. and Vinayak, V. K. 1990. Preliminary evaluation of extracts of *Alstonia scholaris* bark for *in vivo* antimalarial activity in mice. *Journal of Ethnopharmacology* 29(1): 51–57.

Marciano, J., Michailesco, P. and Abadie, M. J. M. (1993) Stereochemical structure characterization of dental gutta-percha. *Journal of Endodontics* 19(1): 31–34.

Maynard, J. 2020. How wounds heal: The 4 main phases of wound healing. Shield Health Care. http://www.shieldhealthcare.com/community/popular/2015/12/18/how-wounds-heal-the-4-main-phases-of-wound-healing/ (accessed 12 January 2020).

McKee, T. C., Fuller, F. W., Covington, C. D. and Cardellina, J. H. 1996. New pyranocoumarins isolated from *Calophyllum lanigerum* and *Calophyllum teysmannii*. *Journal of Natural Products* 59(8): 754–748.

Meena, A. K., Nain Jaspreet, G. N., Meena, R. P. and Rao, M. M. 2011. Review on ethanobotany, phytochemical and pharmacological profile of *Alstonia scholaris*. *International Research Journal of Pharmacy* 2(1): 49–54.

Melecchi, S. S., Martinezb, M. M., Abada, F. C., Zinia, P. P., and Filhoa, I. N. 2002. Chemical composition of *Hibiscus tiliaceus* L. flowers: A study of extraction methods. *Journal of Separation Science* 25(1-2): 86–90.

Meliki, Linda R. and I. Lovadi 2013. Etnobotani tumbuhan obat oleh suku Dayak Iban Desa Tanjung Sari Kecamatan Ketungau Tengah Kabupaten Sintang. *Protobiont* 2(3): 129–135. (abstract in English).

Mendheim, B. 2007. Lost and found: Alice Augusta Ball, an extraordinary woman of Hawai`i Nei. Northwest Hawaii Times, United States.

Misnan, N. M., Omar, M. H. and Wasiman, M. I. 2015. Isolation of clitorin and manghaslin from *Carica papaya* L. leaves by CPC and its quantitative analysis by QNMR. *International Journal of Pharmacological and Pharmaceutical Sciences* 9(9): 37–48.

Mohamed, A. N., Vejayan, J. and Yusoff, M. M. 2015. Review on *Eurycoma longifolia* pharmacological and phytochemical properties. *Journal of Applied Sciences* 15(6): 831–844.

Mohamed, M. R., Mammoud, M. R. and Hayen, H. 2009. Evaluation of antinociceptive and anti-inflammatory activities of a new triterpene saponin from *Bauhinia variegata* leaves. *Zeitschrift fur Naturforschung* 64(11–12): 798–808.

Mohd Ridzuan, M. A. R., Rain, A. N., Zhari, I. and Zakiah, I. 2005. Effect of *Eurycoma longifolia* extract on the Glutathione level in *Plasmodium falciparum* infected erythrocytes *in vitro*. *Tropical Biomedicine* 22(2): 155–163.

Molly, A., Menon, D. B., James, J., Dev, M. S. L., Arun, K., and Thankamani, V. 2011. Phytochemical analysis and antioxidant activity of *Alstonia scholaris*. *Phamacognosy Journal* 3: 13–18.

Mordmuang, A., Shankar, S., Chethanond, U., and Voravuthikunchai, S. P. 2015. Effects of *Rhodomyrtus tomentosa* leaf extract on staphylococcal adhesion and invasion in bovine udder epiderandal tissue model. *Nutrients* 7: 8503–8517.

Moreno, S., Scheyer, T., Romano, C., and Vojnov, A. 2006. Antioxidant and antimicrobial activities of rosemary extracts linked to their polyphenol composition. *Free Radical Research* 40(2): 223–231.

Moura, A. C., Silva, E. L., Fraga, M. C., Wanderley, A. G., Afiatpour, P., *et al.* 2005. Antiinflammatory and chronic toxicity study of the leaves of *Ageratum conyzoides* L. in rats. *Phytomedicine* 12(1–2): 138–142.

Muhameead, R., Hamath, K., Nileena, K., Rosemery, F. and Sinju, K. 2019. Phytochemical Screening of *Justicia gendarussa*. *International Journal of Pharmacognosy and Chinese Medicine* 3(10): 1–6.

Mukhtar, H. M., Ansari, S. H., Ali, M., Naved, T. and Bhat, Z. A. 2004. Effect of water extract of *Psidium guajava* leaves on alloxan-induced diabetic rats. *Pharmazie* 59(9): 734–735.

Murnigsih, T., Subeki, Matsuura, H., Takahashi, K., Yamasaki, M. and Yamato, O. 2005. Evaluation of the inhibitory activities of the extracts of Indonesian traditional medicinal plants against *Plasmodium falciparum* and *Babesia gibsoni*. *Journal of Veterinary Medical Science* 67(8): 829–831.

Muthukumaran J., Vinayagam, R., Ambati, R. R., Xu, B. and Chung, S. S. M. 2018. Guava leaf extract diminishes hyperglycemia and oxidative stress, prevents β cell death, inhibits inflammation, and regulates NF-kB signaling pathway in STZ induced diabetic rats. *Biomedical Research International* 2018(5) Article ID 144601649. doi: 10.1155/2018/4601649. https://www.hindawi.com/journals/bmri/2018/4601649/ (accessed on 4 January 2020).

Nair, J. J. and van Staden, J. 2017. Antifungal activity based studies of Amaryllidaceae plant extracts. *Natural Product Communications* 12(12): 1953–1956.

Nair, S., Kavrekar, V. and Mishra. A. 2013. *In vitro* studies on alpha amylase and alpha glucosidase inhibitory activities of selected plant extracts. *European Journal of Experimental Biology* 3(1): 128–132.

Nair, S., George, J., Kumar, S. and Gracious, N. 2014. Acute oxalate nephropathy following ingestion of *Averrhoa bilimbi* juice. *Case Reports in Nephrology* 2014: 1–5. Article ID 240936, doi.org/10.1155/2014/240936.

Nayak, B. S., Sandiford, S. and Maxwell, A. 2009. Evaluation of the wound-healing activity of ethanolic extract of *Morinda citrifolia* L. leaf. *Evidenced-Based Complementary and Alternative Medicine* 6(3): 351–256.

Nayak, S., Isitor, G. N., Maxwell, A. and Ramdath, D. D. 2007. Wound-healing activity of *Morinda citrifolia* fruit juice on diabetes-induced rats. *Journal of Wound Care* 16(2): 83–86. doi: 10.12968/jowc.2007.16.2.27006.

Nelson, K. M., Dahlin, J. L., Bisson, J., Graham, J., Pauli, G. F., *et al.* 2007. The essential medicinal chemistry of curcumin: Mini perspective. *Journal of Medicinal Chemistry* 60(5): 1620–1637. doi:10.1021/acs.jmedchem.6b00975.

National Institute of Arthritis and Musculoskeletal and Skin Diseases (NIAMS). 2014b. Arthritis and rheumatic diseases. https://www.niams.nih.gov/health-topics/arthritis-and-rheumatic-diseases (accessed on 22 March 2014).

National Institute of Arthritis and Musculoskeletal and Skin Diseases (NIAMS). 2014a. *Living with Arthritis: Health Information Basics for You and Your Family*. https://amac.us/living-arthritis-health-information-basics-family/ (accessed 10 March 2020).

Nguyen-Pouplin, J., Tran, H., Tran, H., Phan, T. A., *et al.* 2007. Antimalarial and cytotoxic activities of ethnopharmacologically selected medicinal plants from South Vietnam. *Journal of Ethnopharmacology* 109(3): 417–427.

Niazi, J., Gupta, V., Chakarborty, P. and Kumar, P. 2010. Anti-inflammatory and antipyretic activity of *Aleurites moluccana* leaves. *Asian Journal of Pharmaceutical and Clinical Research* 3(1): 35–37.

Niljan, J., Jaihan, U., Srichairatanakool, S., Uthaipibull, C. and Somsak, V. 2014. Antimalarial activity of stem extract of *Tinospora crispa* against *Plasmodium berghei* infection in mice. *Journal Health Research* 28(3): 199–204.

Noor, H. and Ashcroft, S. 1989. Antidiabetic effects of *Tinospora crispa* in rats. *Journal of Ethnopharmacology* 27(1–2): 149–161.

Noor, H., and Ashcroft, S. 1998. Pharmacological characterisation of the antihyperglycaemic properties of *Tinospora crispa* extract. *Journal of Ethnopharmacology* 62(1): 7–13.

Noor, H. P., Hammonds, R. S. and Ashcroft, S. J. H. 1989. The hypoglycaemic and insulinotropic activity of *Tinospora crispa*: Studies with human and rat islets and HIT-T15 B cells. *Diabetologia* 32(6): 354–359.

Noorcahyati 2012. Tumbuhan berkhasiat obat etnis asli Kalimantan. Badan Penelitian dan Pengembangan Kehutanan Kementerian Kehutanan. In: Nur Sumedi, Kade Sidiyasa and Faiqotul Falah eds, 1–63. Kartanegara: Balai Penelitian Teknologi Konservasi Sumber Daya Alam ISBN: 978-602-17988-0-5. (In Indonesian).

Nugroho, A., Heryani, H., Choi, J. S..and Park, H. J. 2017. Identification and quantification of flavonoids in *Carica papaya* leaf and peroxynitrite-scavenging activity. *Asian Pacific Journal of Tropical Biomedicine* 7(3): 208–213.

Nurdiana S. and Marziana, N. 2013. Wound healing activities of *Melastoma malabathricum* leaves extraction *Sprague Dawley* rats. *International Journal of Pharmceutical. Science Review and. Research* 20(2): 20–23.

Nurhanan, M. Y., Hawariah, L. P. A., Ilham, A. M. and Shukri, M. A. M., 2005. Cytotoxic effects of the root extracts of *Eurycoma longifolia* Jack. *Phytotherapy Research* 19(11): 994–996.

Nwokocha, C. R., Owu, D. U., Gordon, A., Thaxter. K., McCalla, G., *et al.* 2012. Possible mechanisms of action of the hypotensive effect of *Annona muricata* (soursop) in normotensive Sprague-Dawley rats. *Pharmaceutical Biology* 50(11): 1436–1441. doi: 10.3109/13880209.2012.684690. Epub 2012 Sep 11.

Ogbole, O. O., Akinleye, T. E., Segun, P. A., Faleye, T. C. and Adeniji, A. J. 2018. *In vitro* antiviral activity of twenty-seven medicinal plant extracts from Southwest Nigeria against three serotypes of echoviruses. *Virology Journal* 18 15(1):110–118. doi: 10.1186/s12985-018-1022-7.

Oladejo, O. W., Imosemi, I. O., Osuagwu, F. C., Oluwadara, O. O., Aiku A., *et al.* 2003. Enhancement of cutaneous wound healing by methanolic extracts of *Ageratum conyzoides* in the Wistar rat. *African Journal of Biomedical Research* 6(1): 27–31.

Omar, S., Zhang, J., MacKinnon, S., Leaman, D., Durst, T. and Philogene, B. J. R. 2003. Traditionally-used antimalarials from the Meliaceae. *Current Tropics in Medicinal Chemistry* 3(2): 133–139.

Omar, S. N. C., Abdullah, J. O., Khairoji, K. A., Chin, S. C., and Hamid, M. 2013. Effects of flower and fruit extracts of *Melastoma malabathricum* Linn. on growth of pathogenic bacteria: *Listeria monocytogenes, Staphylococcus aureus, Escherichia coli*, and *Salmonella typhimurium. Evidence-Based Complementary and Alternative Medicine* 2013: 459089–459100.

Ompusunggu, S. 2015. Malaria hutan di Provinsi Kalimantan Tengah dan Kalimantan Selatan, Indonesia Tahun 2013. *Jurnal Ekologi Kesehatan* 14(2): 145–156.

Padmanaban, S., Pandian, L., Kalavathi, M. K., Swetha, R., Anitha, P., *et al.* 2013. Flavonoids from *Carica papaya* inhibits NS2BNS3 protease and prevents Dengue 2 viral assembly. *Bioinformation* 9(18): 889–895.

Palu, A., Su, C., Zhou, B. N., West, B. and Jensen, J. 2010. Wound healing effects of noni (*Morinda citrifolia* L.) leaves: A mechanism involving its PDGF/A2A receptor ligand binding and promotion of wound closure. *Phytotheraphy Research* 24(10): 1437–1441. doi: 10.1002/ptr.3150.

Pappas, S. 2018. 12th-Century Shipwreck Came with Handy 'Made in China' Tag. https://www.livescience.com/62588-ancient-shipwreck-made-in-china.html (accessed 20 March 2020).

Paramapojna, S., Ganzera, M., Gritsanapana, W., and Stuppner, H. 2008. Analysis of naphthoquinone derivatives in the Asian medicinal plant Eleutherine americana by RP-HPLC and LC–MS. *Journal of Pharmaceutical and Biomedical Analysis* 47: 990–993.

OrthoInfo. 2007. Arthritis: An Overview. https://orthoinfo.aaos.org/en/diseases--conditions/arthritis-an-overview/ (accessed October 2007).

Parasuraman, S., Ching, T. H., Leong, C. H. and Banik, U. 2019. Antidiabetic and anti-hyperlipidermic effects of a methanolic extract of *Mimosa pudica* (Fabaceae) in diabetic rats. *Egyptian Journal of Basic and Applied Sciences* 6(1): 137–148.

Park, S., Nguyen, X. N., Phan, V. K. and Chau, V. M. 2014. Five new quassinoids and cytotoxic constituents from the roots of *Eurycoma longifolia*. *Bioorganic & Medicinal Chemistry Letters* 24(16): 3835–3840.

Parkavi, M., Vignesh, K., Selvakumar, J., Mohamed, M. and Ruby, J. J. 2012. Antibacterial activity of aerial parts of *Imperata cylindrica* (L) Beauv. *International Journal of Pharmaceutical Sciences and Drug Research* 4(3): 209–212.

Patel, S. and Adhav, M. 2016. Comparative phytochemical screening of ethanolic extracts (flower and leaf) of morphotypes of *Hibiscus rosa-sinensis* L. *Journal of Pharmacognosy and Phytochemistry* 5(3): 93–95.

Pavithra, P. S., Janan, I. V. S., Charumathi, K. H., Indumathy, R., Sirisha. P. *et al.* 2010. Antibacterial activity of plants used in Indian herbal medicine. *International Journal of Green Pharmacy* 2010: 22–28.

Phongpaichit, S., Pujenjob, N., Rukachaisirikul, V. and Ongsakul, M. 2004. Antifungal activity from leaf extracts of *Cassia alata* L., *Cassia fistula* L. and *Cassia tora* L. Songklanakarin. *Journal of Science and Technology* 26(5): 741–748.

Pilerood, S. A. and Prakash, J. 2011. Chemical composition and antioxidant properties of ginger root (*Zingiber officinale*). *Journal of Medicinal Plant Research* 4(24): 2674–2679.

Ponnusamy, Y. 2016. *In vitro* antibacterial and wound healing properties of a standardized polyphenols-rich fraction of *Dicranopteris linearis* (Burm.) Underwood. PhD diss., Universiti Sains Malaysia, Malaysia.

Ponou, B. K., Teponno, R. B., Tapondjou, L. A., Barboni, L., Lacaille-Dubois, M.-A. *et al.* 2019. Steroidal saponins from the aerial parts of *Cordyline fruticosa* L. var. *strawberries*. *Fitoterapia* 134: 454–458. doi: 10.1016/j.fitote.2019.03.019.

Pradityo, T., Santoso, N. and Zuhud, E. A. 2016. Etnobotani kebun Tembawang suku Dayak Iban, Desa Sungai Mawang, Kalimantan Barat. *Media Konservasi* 21(2): 183–198. (In Indonesian).

Prasad, B. D., Kanth, B. C., Babu, R., Kumar, K. P. and Sastry, V. G. 2011. Screening of wound healing activity of bark of *Aleurites moluccana*. *International Journal of Pharmaceutical Research and Analysis* 1(1): 21–25.

Prasenjit, M., Tanaya, G., Gupta, S., Basudeb, B. and Kumar, M. P. (2016). Isolation and characterization of a compound from the leaves of *Cassia alata* Linn. *E-Cronicon*. https://pdfs.semanticscholar.org/faca/6e8e2ebbaa4b4abb0a215276ed4a8a32d4a7.pdf (accessed on 22 July 2020).

Prastiwi, R., Siska, E. B. U., and Witji, G. P. 2016b. Antihypertensive and diuretic effects of the ethanol extract of *Colocasia esculenta* (L.) Schott. leaves. *Ilmu Kefarmasian Indonesia* 14(1): 99–102. (abstract in English).

Pratyush, K., Misra, C. S., James, J., Lipin, M. S., Dev, A., *et al.* 2011. Ethnobotanical and pharmacological study of *Alstonia* (Apocynaceae) – a review. *Journal of Pharmaceutical Sciences and Research* 3(8): 1394–1403.

Prescott, T. A. K., Homot, P., Lundy, F. T., Patrick, S., Cámara- Leret, R., *et al.* 2017. Tropical ulcer plant treatments used by Papua New Guinea's Apsokok nomads: Fibroblast stimulation, MMP protease inhibition and antibacterial activity. *Journal of Ethnopharmacology* 205: 1–260 doi.org/10.1016/j.jep.2017.05.001.

Qaddoori, A. G. 2016. Antimicrobial evaluation of selected medicinal plants using molecular approach. PhD diss., University of Salford Manchester, United Kingdom.

Radhakrishnan, N., Lam, K. W. and Norhaizan, M. E. 2017. Molecular docking analysis of *Carica papaya* L. constituents as antiviral agent. *International Food Research Journal* 24(4): 1819–1825.

Rady, I., M. B. Bloch, Roxane-Cherille N. Chamcheu, S. B. Mbeumi, *et al.* 2018. Anticancer properties of Graviola (*Annona muricata*): A comprehensive mechanistic review. *Oxidative Medicine and Cellular Longevity* 2018: 1–39.

Ragasa, C. Y., Soriano, G., Torres, O. B., Don, M. J. and Shen, C. C. 2012. Acetogenins from *Annona muricata*. *Pharmacognosy Journal* 4(32): 32–37.

Rahman, N. N. N. A., Furuta, T., Kojima,S., Takane, K. and Mustafa, A. M. 1999. Antimalarial activity of extracts of Malaysian medicinal plants. *Journal of Ethnopharmacology* 64(3): 249–254.

Raja, D. P., Manickam, V. S., De Britto, A. J., Gopalakrishnan, S., Ushioda, T. 1995. Chemical and chemotaxonomical studies on *Dicranopteris* species. *Chemical and Pharmaceutical Bulletin* 43(10): 1800–1803.

Rajasekaran, D., Palombo, E. A., Yeo, T. C., Lim, S. L. D., and Tu, C. L. 2013. Identification of traditional medicinal plant extracts with novel antiinfluenza activity. *Plos One.* doi: 10.1371/journal.pone.0079293. https://journals.plos.org/plosone/article?id=10.1371/journal.pone.0079293 (accessed on 22 July 2020).

Ranasinghe, P., Ranasinghe, P., Abeysekera, W. P., Premakumara, G. A., Perera, Y. S., *et al.* (2012). *In vitro* erythrocyte membrane stabilization properties of *Carica papaya* L. leaf extracts. *Pharmacognosy Research* 4(4): 196–202. doi: 10.4103/0974-8490.102261.

Rehman, S. U., Choe, K. and Yoo, H. H. 2016. Review on a traditional herbal medicine, *Eurycoma longifolia* Jack (Tongkat Ali): its traditional uses, chemistry, evidence-based pharmacology and toxicology. *Molecules* 21(3): 1–31.

Rocha, M. F. G., Sales, J. A., da Rocha, M. G., Gladino, M. G., Gladino, L. M., de Aguiar, L., *et al.* 2019. Antifungal effects of the flavonoids kaempferol and quercetin: a possible alternative for the control of fungal biofilms. *The Journal of Bioadhesion and Biofilm Research* 35(3): 320–328.

Rosidah, Y. M. F., Sadikun, A., Ahmad, M., Akowuah, G. A. and Asmawi, M. Z. 2009. Toxicology evaluation of standardized methanol extract of *Gynura procumbens*. *Journal of Ethnopharmacology* 123(2): 244–249. doi: 10.1016/j.jep.2009.03.011.

Rosli, N., Sumathy, V., Vikneswaran, M. and Sreeramanan, S. 2014. Growth profile and SEM analyses of *Candida albicans* and *Escherichia coli* with *Hymenocallis littoralis* (Jacq.) Salisb leaf extract. *Tropical Biomedicine* 31(4): 871–879.

Ruan, C.-T., Lam, S.-H., Chi, T.-C., Lee, S.-S. and Su, M.-J. 2012. Borapetoside C from *Tinospora crispa* improves insulin sensitivity in diabetic mice. *Phytomedicine* 19(8–9): 719–724.

Rufina, S. D. and Reni M. 2013. Etnobotani tumbuhan obat suku Dayak Pesaguan dan implementasinya dalam pembuatan flash card biodiversitas. Program Studi Pendidikan Biologi FKIP, Universitas Tanjungpura, Pontianak. (In Indonesian).

Rungruang, T. and Thidarut B. 2009. *In vivo* antiparasitic activity of the Thai traditional medicine plant – *Tinospora crispa* – against *Plasmodium yoelii*. *Southeast Asian Journal of Tropical Medicine and Public Health* 40(5): 898–900.

Sabirin, I. P. R. and Yuslianti, E. R. 2016. Benefits of ethanol based Noni leaf (*Morinda citrifolia* L.) extract on oral mucosal wound healing by examination of fibroblast cells. *Journal of Dentistry Indonesia* 23(3): 59–63.

Saewan, N., Sutherland, J. D. and Chantrapromma, K. 2006. Antimalarial tetranortriterpenoids from the seeds of *Lansium domesticum* Corr. *Phytochemistry* 67(20): 2288–2293.

Saising, J., Ongsakul, M. and Voravuthikunchai, S. P. 2011. *Rhodomyrtus tomentosa* (Aiton) Hassk. Ethanol extract and rhodomyrtone: A potential strategy forthe treatment of biofilm-forming staphylococci. *Journal of Medical Microbiology* 60: 1793–1800.

Samy, J., Sugumaran, M. and Lee, K. L. W. 2014. 100 Useful herbs of Malaysia and Singapore: An introduction to their medicinal, culinary, aromatic and cosmetic uses. Marshall Cevendish Editions, Singapore.

Santoso, E. A., Jumari and Utami, S. 2019. Inventory and biodiversity of medicinal plants of Dayak Tomun society in Lopus village Lamadau regency, Central Kalimantan. The 8th International Seminar on New Paradigm and Innovation on Natural Science and Its Application 26 September 2019, Central Java, Indonesia.

Sari, A., Linda, R. and Lovadi, I. 2015. Pemanfaatan tumbuhan obat pada masyarakat suku Dayak Jangkang Tanjung di Desa Ribau Kecamatan Kapuas Kabupaten Sanggau. *Protobiont* 4(2): 1–8. (In Indonesian).

Sari, R. Y., Wardenaar, E. and Muflihati 2014. Etnobotani tumbuhan obat di dusun Serambai Kecamatan Kembayan Kabupaten Sanggu, Kalimantan Barat. *Jurnal Hutan Lestari* 2(3): 397–387. (In Indonesian).

Sellato, B. 2002. Non-timber forest products and trade in eastern Borneo. *Bois Et Forêts des Tropiques* 271(1): 37–50.

Sengupta, A. and Gupta, J. K. 1973. The component fatty acids of chaulmoogra oil. *Journal of the Science of Food Agriculture* 24: 669–674.

Setyowati, F. M. 2010. Etnofarmakologi dan pemakaian tanaman obat suku Dayak Tunjung di Kalimantan Timur. *Media Penelitian dan Pengembangan Kesehatan* 20(3): 104–112 (abstract in English).

Setyowati, F. M., Soedarsono R. and Siti S. 2005 Etnobotani masyarakat Dayak Ngaju di Daerah Timpah Kalimantan Tengah. *Jurnal Teknologi Lingkungan* 6(3): 104–112, 502–510 (abstract in English).

Shamili, G. and Santhi, G. 2019. Identification and characterization of bioactive compounds of leaves of *Justicia gendarussa* Burm. *f. International Journal of Scientific Research in Biological Sciences* 6(1): 145–153.

Sharma, P. K., Singh, V. and Ali, M. 2016. Chemical composition and antimicrobial activity of fresh rhizomes essential oil of *Zingiber officinale* Roscoe. *Pharmacognosy Journal* 8(3): 185–190.

Shi, L. S., Kuo, S. C., Sun, H. D., Morris-Natschke, S. L., Lee, K. H., *et al.* 2014. Cytotoxic cardiac glycosides and coumarins from *Antiaris toxicaria. Bioorganic Medicinal Chemistry* 22(6): 1889–1898. doi: 10.1016/j.bmc.2014.01.052. Epub 2014 Feb 6.

Simpkin, A. 1928. The treatment of leprosy. *British Journal of Nursing* 1928: 313–314.

Singh, D. and Chaudhur, P. K. 2018. Structural characteristics, bioavailability and cardio-protective potential of saponins. *Integrative Medicine Research* 7(1): 33–43. doi: 10.1016/j.imr.2018.01.003.

Singh, G. and Saxena, R. K. 2016. Evaluation of antimicrobial activity of *Tinospora cordifolia* and *Hymenocallis littoralis* medicinal plants by using different solvents extract. *International Research Journal of Engineering and Technology* 3(3): 928–931.

Sinnathambi, A., Mazumder, P. M., Lohidasan, S. and Thakurdesai, P. 2010. Antidiabetic and antihyperlipidemic activity of leaves of *Alstonia scholaris* Linn. R. Br. *European Journal of Integrative Medicine* 2(1): 23–32.

Sirirak, T. and Voravuthikunchai, S. P. 2011. *Eleutherine americana*: A candidate for the control of *Campylobacter* species. *Poultry Science* 90: 791–796.

Siti Fauziah, Y. 2013. Ensiklopedia Tumbuhan Ubatan Malaysia. Selangor: AR-Risalah Product Sdn.Bhd. (In Malay).

Sivashanmugam, A. T. and Chatterjee, T. K. 2012. Anticataractogenesis activity of Polyalthia *longifolia leaves* extracts against glucose-induced cataractogenesis using goat lenses *in vitro. European Journal of Experimental Biology* 2(1): 105–113.

Skornickova, J. L. and Gallick, D. 2010. Singapore botanic gardens pictorial pocket guide 2: The Ginger Garden. Singapore: National Parks Board Singapore Botanic Gardens.

Soni, D. and Gupta, A. 2011. An evaluation of anti-pyretic and analgesic potentials of aqueous root extract of *Hibiscus rosa-sinensis* Linn. (Malvaceae). *International Journal of Research in Phytochemistry and Pharmacology* 1(3): 184–186.

Srisuwan, S., Mackin, K. E., Hocking, D., Lyras, D., Bennett-Wood, V., *et al.* 2018. Antibacterial activity of rhodomyrtone on *Clostridium difficile* vegetative cells and spores *in vitro*. *International Journal of Antimicrobial Agents* 52: 724–729.

Su, B. N., Pawlus, A. D., Jung, H. A., Keller, W. J., McLaughlin, J. L., *et al.*. 2005. Chemical constituents of the fruits of *Morinda citrifolia* (Noni) and their antioxidant activity. *Journal Natural Products* 68(4): 592–595.

Subramanian R., Marhain, N., Bhattacharjee, A. and Ahmad, S. 2018. Evaluation of free radical scavenging, anti-inflammatory and wound healing effects of *Nephelium lappaceum* leaf extract *Pharmacologyonline* 2: 93–100. https://pharmacologyonline.silae.it/files/archives/2018/vol2/PhOL_2018_2_A010_Subramanian.pdf (accessed 26 January 2020).

Sudarmono. 2018. Biodiversity of medicinal plants at Sambas Botanical Garden, West Kalmantan, Indonesia. *The Journal of Tropical Life Science* (92): 116–122. doi: 10. 11594/jtis.08.02.04.

Sulaiman, M. R., Zakaria, Z. A., Daud, I. A., Ng, F. N., Ng, Y. C. and Hidayat, M. T. 2008. Antiociceptive and anti-inflammatory activities of the aqueous extract of Kaempferia galanga leaves in animal models. *Journal of Natural Medicines* 62(2): 221–227.

Sule,W. F., Okonko, I. O., Joseph, T. A., Ojezele, M. O., Nwanze, J. C., *et al.* 2010. *In vitro* antifungal activity of *Senna alata* Linn. crude leaf extract. *Research Journal of Biological Sciences* 5(3): 275–284.

Sundarasekar, J., Sahgal, G. and Subramaniam, S. 2012. Anti-candida activity by *Hymenocallis littoralis* extracts for opportunistic oral and genital infection *Candida albicans*. *Bangladesh Journal of Pharmacology* 7: 211–216.

Sunilson, A., James, J., Thomas, J., Paulraj, J., Rajavel, V., *et al.* 2008. Antibacterial and wound healing activities of *Melastoma malabathricum L. African Journal of Infectious Diseases* 2(2): 68–73.

Supiandi, M. I., Mahanal, S., Zubaidah, S., Julung, H. and Ege, B. 2019. Ethnobotany of traditional medicinal plants usedby Dayak Desa Community in Sintang, West Kalimantan, Indonesia. *Bodiversitas Journal of Biological Diversity* 20(5): 1264–1270.

Sutthammikorn, N., Supajatura, V., Niyonsaba, F., Nakano, N., Okumura, K., *et al.* 2018. Evaluation of wound healing efficacy of *Gynura procumbens* leaf extract in streptozotocin-induced diabetic mice. 4th Global Summit on Herbals and Traditional Medicine. 3–4 October, 2018. Osaka, Japan.

Takoy, D. M., Linda, R. and Lovadi, I. 2013. Tumbuhan berkhasiat obat Suku Dayak Seberuang di kawasan hutan Desa Ensabang Kecamatan Sepauk Kabupaten Sintang. *Protobiont* 2(3): 122–128. (In Indonesian).

Tang, C.-S. (1979). New macrocyclic, Δ1-piperideine alkaloids from papaya leaves: dehydrocarpaine I and II. *Phytochemistry* 18(4): 651–652.

Tarkang, P. A., Okalebo, Siminyu, J. D., Ngugi, W. N., Mwara, A. M., *et al.* 2015. Pharmacological evidence for the folk use of Nefang: Antipyretic, anti-inflammatory and antinociceptive activities of its constituent plants. *BMC Complementary and Alternative Medicine* 15(1): 1–11.

Tenga, W.-C., Chana, W., Suwanarusk, R., Ong, A., Ho, H. K., Russell, B. 2019. *In vitro* antimalarial evaluations and cytotoxicity investigations of *Carica papaya* leaves and carpaine. *Natural Products Communications* 14(1): 33–36.

Thanh, T. B., Thanh, H. N., Thi Ly, H. D., Huong, L.-T.-T., Loi, V. D., *et al.* 2015. Flavonoids from leaves of *Tetracera scandens* L. *Journal of Chemical and Pharmaceutical Research* 7(3): 2123–2126.

Thitilertdecha, N., Teerawutgulrag, A., Kilburn, J. D. and Rakariyatham, N. 2010. Identification of major phenolic compounds from *Nephelium lappaceum* L. and their antioxidant activities. *Molecules* 15(3): 1453–1465. doi: 10.3390/molecules15031453.

Umar, M. I., Asmawi, M. Z., Sadikun, A., Atangwho, I. J. and Yam, M. F. 2012. Bioactivity guided isolation of ethyl-*p*-methoxycinnamate, an anti-inflammatory constituent, from *Kaempferia galanga* L. extract. *Molecules* 17: 8720–8734.

Uphof, J. C. Th. 1959. Dictionary of economic plants. Weinheim: Lubrecht and Cramer Ltd.

Vasant, O. K., Vijay, B. G., Virbhadrappa, S. R., Dilip, N. T., Ramahari, M. V., *et al.* 2012. Antihypertensive and diuretic effects of the aqueous extract of *Colocasia esculenta* Linn. leaves in experimental paradigms. *Iran Journal of Pharmaceutical Research* 11(2): 621–634.

Vishwanath, V. and Rao, H. M. 2019. Gutta-percha in endodontics – A comprehensive review of material science. *Journal of Conservative Dentistry* 22(3): 216–222.

Walid, E. and Awad, M. 2015. HPLC Analysis of quercetin and antimicrobial activity of comparative methanol extracts of *Shinus molle* L. *Biology* 4(11): 550–558.

Wan, J. Y., Gong, X., Jiang, R., Zhang, Z. and Zhang, L. 2012. Antipyrectic and anti-inflammatory effects of asiaticoside in lipolysaccharidep\treated rat through up-regulation of heme oxygenase 1. *Phytoteraphy Research* 27(8): 1136–1142.

Wang, L., Wang, B., Li, H., Lu, H., Qiu, F., *et al.* 2012. Quercetin, a flavonoid with anti-inflammatory activity, suppresses the development of abdominal aortic aneurysms in mice. *European Journal of Pharmacology* 690(1–3): 133–141.

Wang, X., Hu, C., Chen, Ai Q., Wang, Z., *et al.* 2015. Isolation and identification of carpaine in *Carica papaya* L. leaf by HPLC-UV method. *International Journal of Food Properties* 18(7): 1505–1512. doi: 10.1080/10942912.2014.900785.

Wang, Y. H. and Yu, X. Y. 2018. Biological activities and chemical composition of volatile oil and essential oil from the leaves of *Blumea balsamifera*. *Journal of Essential Oil Bearing Plants* 26(6): 1511–1531.

Wernsdorfer, W. H., Sabariah, I., Chan, K. L., Congpuong, K. and Gunther Wernsdorfer, G. 2009. Activity of *Eurycoma longifolia* root extract against *Plasmodium falciparum in vitro*. *Wiener Klinische Wochenschrift* 121(3): 23–26.

Wisutthathum, S., Kamkaew, N., Chatturong, A., U., Paracha, U., KornkanokIngkaninan, *et al.* 2019. Extract of *Aquilaria crassna* leaves and mangiferin are vasodilators while showing no cytotoxicity. *Journal of Traditional and Complementary Medicine* 9(4): 237–242.

Wong, K. C., Ali, D. M. H. and Boey, P.-L. 2012. Chemical constituents and antibacterial activity of *Melastoma malabathricum* L. *Natural Product Research* 26(7): 609–618.

Wright, C. W., Allen, J. D., Phillipson, J. D., Kirby, G. C. and Warhurst, D. C. 1993. *Alstonia* species: Are they effective in malaria treatment? *Journal of Ethnopharmacology* 40: 41–45.

Wuthi-udomlert, M., Prathanturarug, S. and Soonthornchareonnon, N. 2003. Antifungal activities of *Senna alata* extracts using different methods of extraction. Proceedings of the International Conference on MAP. *Acta Horticuturae* 59: 205–208.

Wuthi-udomlert, M., Kupittayanant, P. and Gritsanapan, W. 2010. *In vitro* evaluation of antifungal activity of anthraquinone derivatives of *Senna alata*. *Journal of Health Research* 24(3): 117–122.

World Health Organization. 1998. Medicinal plants in South Pacific. WHO Regional Publications, Manila. (accessed on 28 July 2020).

World Health Organization. 2017a. HIV/AIDS. http://www.who.int/mediacentre/factsheets/fs360/en/ (accessed on 7 January, 2020).

World Health Organisation. 2017b. New Hepatitis Data Highlight Need for Urgent Global Response. http://www.who.int/mediacentre/news/releases/2017/global-hepatitis-report/en/ (accessed 7 January 2020).

World Health Organisation. 2018. Diabetes. http://www.who.int/en/newsroom/fact-sheets/detail/diabetes (accessed 5 November 2018).

Xie, Y., Yang, W., Chen, X., and Ren, L. 2014. Antibacterial activities of flavonoids: Structure-activity relationship and mechanism. *Current Medicinal Chemistry* 22(1): 132–149.

Xu, Y., Niu, Y., Gao, Y., Wang, F., Qin, W., *et al.* 2017. Borapetoside E, a clerodane diterpenoid extracted from *Tinospora crispa*, improves hyperglycemia and hyperlipidemia in high-fat-diet-induced Type 2 diabetes mice. *Journal of Natural Products* 80(8): 2319–2327.

Xu, Z.-Q., Bar row, W., Suling, W. J., Westbrook, L. and Barrow, E. 2004. Anti-HIV natural product (+)-calanolide A is active against both drug-susceptible and drug-resistant strains of *Mycobacterium tuberculosis*. *Bioorganic & Medicinal Chemistry* 12(5): 1199–1207.

Yadav, P., Choudhury, S., Barua, S., Khandelwal, N., Kumar, N., *et al.* 2019. *Polyalthia longifolia* leaves methanolic extract targets entry and budding of viruses – an *in vitro* experimental study against paramyxoviruses. *Ethnopharmacology* 7(248): 112279. doi: 10.1016/j.jep.2019.112279. [Epub ahead of print]

Yajid, A. I., Ab Rahman, H. S., Wong, M. P. K. and Zain, W. Z. W. 2018. Potential benefits of *Annona muricata* in combating cancer: A review. *The Malaysian Journal of Medical Sciences* 25(1): 5–15.

Yang X., Wang J., and Xu Y. 2015. A traditional Chinese medicine preparation for treating damp heat toxin accumulation type chronic skin ulcer and its production method. CN. Patent No. 104666446. State Intellectual Property Office of the P.R.C., Beijing.

Yao, L. J., Chiew, C. H. and Zakaria, N. A. 2019. The medicinal uses, toxicities and anti-inflammatory activity of Polyalthia species (Annonaceae). *Journal of Ethnopharmacology* 229(30): 303–325.

Yapp, D. T. T. and Yap, S. Y. 2003. *Lansium domesticum*: skin and leaf extracts of this fruit tree interrupt the lifecycle of *Plasmodium falciparum* and are active towards a chloroquine resistant strain of the parasite (T9) *in vitro*. *Journal of Ethnopharmacology* 85(1): 145–150.

Yazan, L. S., Ong, Y. S., Zaaba, N. E., Ali, R. M., Foo, J. B., *et al.* 2015. Anti-breast cancer properties and toxicity of *Dillenia suffruticosa* root aqueous extract in BALB/c mice. *Asian Pacific Journal of Tropical Biomedicine* 5(12): 1018–1026.

Yoopan, N., Thisoda, P., Rangkadilok, N., Sahasitiwat, S.,Pholphana, N. 2007. Cardiovascular effects of 14-deoxy-11,12-didehydroandrographolide and *Andrographis paniculata* extracts. *Planta Medica* 73(6): 503–511.

Younas, N. and Akhtar, T. 2014. Effect of *Carica papaya* leaf formulation on hematology and serology of normal rat. *Biologia (Pakistan)* 60(1): 139–142.

Yu, H. F., Huang, W. Y., Ding, C. F., Wei, X., Zhang, L.-C., Qin, *et al.* 2018. Cage-like monoterpenoid indole alkaloids with antimicrobial activity from *Alstonia scholaris*. *Tetrahedron Letters* 59: 2975–2978.

Yusro, F., Mariani, Y., Diba, F. and Ohtani, K. 2014. Inventory of medicinal plamts for fever used by four Dayak sub ethnic in West Kalimantan, Indonesia. *Kuroshio Science* 8(1): 33–38.

Yusuf, H., Mustofa Wijayanti, M., Asmah Susidarti, A. R., and Budi Setia Asih, P. 2013. A new quassinoid of four isolated compounds from extract *Eurycoma longifolia*, Jack roots and their *in-vitro* antimalarial activity. *International Journal of Research in Pharmaceutical and Biomedical Sciences* 4(3): 728–734.

Zahra, A. A., Kadir, F. A., Mahmood, A. A., Hadi, A. A. A., Suzy, S. M., *et al.* 2011. Acute toxicity study and wound healing potential of *Gynura procumbens* leaf extract in rats. *Journal of Medicinal Plants Research* 15(12): 2551–2558.

Zakaria, Z. and Vikneswaran, M. 2010. Antimicrobial activity of *Dicranopteris linearis* (Glecheniaceae) (resam) extracts prepared via maceration and Soxhlet method. International Environmental and Health Conference, 6–20 July 2012, Universiti Sains Malaysia, Penang, Malaysia.

Zakaria, Z. A., Fasya, Z., Nor, R., Gopalan, H., Sulaiman, M. R. and Abdullah, F. 2006. Antinociceptive and anti-inflammatory activities of *Dicranopteris linearis* leaves chloroform extract in experimental animals. *Yakugaku zasshi: Journal of the Pharmaceutical Society of Japan* 126: 1197–1203. doi: 10.1248/yakushi.126.1197.

Zarta, A. R. Farida, A., Wiwin, S., Irawan, W. K. and Enos, T. A. 2018. Identification and evaluation of bioactivity in forest plants used for medicinal purposes by the Kutai community of East Kalimantan Indonesia. *Biodiversitas Journal of Biological Diversity* 19(1): 253–259.

Zgoda, J. R., Freyer, A. J., Killmer, L. B. and Porter, J. R. 2001. Polyacetylene carboxylic acids from Mitrephora celebica. *Journal of Natural Products* 64(10): 1348–1349. doi.org/10.1021/np0102509.

Zhang, C. Y. and Tan, B. K. 1996. Hypotensive activity of aqueous extract of *Andrographis paniculata* in rats. *Clinical and Experimental Pharmacology and Physiology* 23(8): 675–678. doi: 10.1111/j.1440-1681.1996.tb01756.x.

Zhengxiong, C., Huizhu, H., Chengruui, W., Yuhui, L., Jianmi, D. et al. 1986. Hongconin, a new naphthalene derivative from Hong-Cong, the rhizome of *Eleutherine americana* Merr. and Heyne (Iridaceae). *Chemical and Pharmaceutical Bulletin (Tokyo)* 34: 2743–2746.

# Index

Note: Page locators in italics indicate figures and those in bold indicate tables.

9781138601079